Wild Old Woman

Wild Old Woman:

*A Meta-Memoir from
Burning Man to Bhutan*

Joan Maloof

© 2024 Joan Maloof
www.JoanMaloof.com

All rights reserved.

ISBN 979-8-218-32433-9

Cover: the author gesturing toward the Tiger's Nest monastery, Bhutan.

Categories:
1. Biography & Autobiography > Adventurers & Explorers > Women > Aging
2. Biography & Memoirs > Arts & Literature > Authors
3. Travel > Travel Writing > Specialty Travel
4. Self Help > Personal Transformation

For the Forests Publishing

Wayfarer, there is no way,
you make the way as you go.
There is the path you make,
and nothing more.
When you look back
you see only footprints
on a road you will never walk again.
There is no way,
only foam trails on the sea.

by Antonio Machado, translated by J. Maloof

Table of Contents

Preface 1
Zipolite One 3
Burning Man One (age 53) 15
The Human Ecosystem 41
Losing Rick 47
The Forests 67
Burning Man Two (age 55) 71
Zipolite Two 93
People are So Generous 101
Love/Hate/Love….? 105
Burning Man Three (age 63) 121
Zipolite Three 135
Bhutan on my Mind 147
Bhutan for Real 158
Meta One 185
Meta Two 193
Post Script 202
Questions for Reading Groups 205

Preface

This story covers parts of my forties, fifties, and sixties. I will start in my forties, when I was a scientist, a teacher, a writer, and a 'householder' – the type with pets and gardens and sourdough starter. By the end I am something entirely different.

Stories may be told in many ways, and a telling in simple chronological order is not always the most compelling, but a main theme of this story is aging – what it feels like to be a woman on the verge of *the change,* and then over it, and then becoming a woman *of a certain age,* and then finally facing elderhood. This arrow goes in one direction only, and that is primarily how this story is told. As the 17th century Japanese pilgrim Basho says, "The years that come and go are travelers too."[1]

From this distance I can also relate to his line, "There will be hardships enough to turn my hair white, but I shall see with my own eyes places about which I have only heard!"

Yesterday I was interviewed for a podcast and the first question was, "When did your passion for nature begin and how did that turn into a career?"

I answered, "at birth," and "I have turned it into many careers." I was born with an affinity for plants, and I had turned that into a life's work of research and teaching. I may have taught you biology or plant taxonomy or environmental studies. You may have been on my tenure committee. You might never have guessed what was beneath that modestly-dressed professorly façade. But I was growing and changing too, just like the plants I studied. The container I had built so carefully, with advanced degrees and grant funding, could contain me no longer. I had to get *wilder* if I were going to save the forests that were falling all around me. I had to get *wilder* if I were going to learn all the lessons meant for me in this lifetime. I had to stay curious, I could not be timid, I had to be like the wildland I defended.

Everything in this story is true. Names of dead people and public figures have been retained, but names of living people and their identifying details have been changed.

Zipolite One

When I turn back I see only footprints on a road I will never walk again... It was a sweet time. I was in love. I had no idea what was ahead on that path, what changes I would go through, what grief and joy lay waiting there.

I look over at Rick, my husband of decades, across the wide, white, king-sized bed. His skin is dark against the sheets; he tans so easily. I am much lighter, but still about as dark as I ever get, and my long hair has developed golden highlights. At this hour there is a chill in the air, but it will soon be hot – a welcome change from the bone-chilling winter weather at our home on the east coast of the US.

We have slept with the shutters open, as usual. No glass, no screens, just simple wooden shutters that can be swung closed in case of a storm. I like to leave them open so I can be woken by the changing color of the sky. But something else has woken me this morning, it sounds like a pack of puppies, but it's not puppies -- it's parrots! A flock of green and yellow parrots are in the tree just outside the window.

I slip on a sundress and head to the kitchen to pump water from a big plastic jug into a small metal stovetop espresso maker. I open the valve on the propane tank, then turn the knob on the stove's burner while holding a lit match over it. The whoosh of flame makes me jump. I wait patiently, watching the birds, until I see steam coming from the spout, then I know it is ready to pour. I bring us each a cup and we sip in silence watching the sky turn various colors until, finally, the orange disc of the sun rises above the horizon.

A hummingbird dances through the air visiting the purple bougainvillea flowers. I don't know the exact name of the other shrub, but its many small leaflets and long lumpy pods tell me it is in the pea family. I am here, I am completely aware, I am in gratitude, I am in love.

The day unfolds like so many others on the Pacific coast of Mexico. There will be breakfast, either at our place or at a little café. There will be a walk down the beach. There will be a swim. There will be lazy hours in the middle of the day, sitting in the shade, alternating between reading a book or staring off into the sea... maybe looking for whales, or watching the pelicans fly in formation down the ridge of the breaking waves. Today I am reading

Nature, Man and Woman by Alan Watts, written in 1958, two years after I was born. I'm forty-something and Rick is in his fifties. Our love story would take a whole book of its own to tell. Few believed that we would last, but we have, and after all these years together we are perfectly in sync. In the afternoon we hike back to our casita to shower and change.

I am on winter break from the university where I teach. At home I am so careful with my reputation that I am uncomfortable even having a drink at the local bar, but here in Mexico I feel free, free, free.

Before the break there was the usual countdown in the mailroom: "only three weeks to go," then, "only two weeks to go," then "almost there." All of us wishing time away, wishing our lives away. I recognized the disease, and I wanted out of it. I wanted a healthy relationship with time.

Time. I put so much time into that other life already... Making sure to join clubs and win honors in high school so I could get into college; doing well in college so I could go to grad school. Taking the "important" jobs in the summers instead of the ones that were most fun. Always building that resume. Missing out on parties and movies so I could study for my statistics exams -- memorizing the formula for 'standard deviation' and understanding *p*-values.

Then finally the Ph.D. -- almost five years for that stretch. Following that came the hours and hours in the classroom teaching, until finally, our society's holy grail, the letter awarding me tenure. That is what I am escaping now -- even if it is only for a few weeks.

Toward the end of the day, after my shower, I smooth my skin with coconut oil, then put on a pale-yellow cotton dress and my necklace with the white whale-tail carved from bone. I slide bangles on my wrists and fasten the beaded anklet that has the tiny bells. When Rick is ready to walk back to the beach with me, I lift a small cloth bag over my shoulder. Inside it is my favorite sarong, my sunglasses, and enough pesos to get me through the evening. I slip my toes into my flip-flops, and we head down the hill through the shady gardens.

After a ten-minute walk we reach the beach and I spread my sarong on the sand. We are in a good spot to watch the sunset – a nightly ritual for both residents and visitors. Even the dogs gather on the beach this time of day, they run free and greet each other like the old friends that they are. Children jump over the white line of surf as it reaches its zenith. All of this brings me joy, and I smile. There is nowhere else I would rather be.

But sunset will not happen for another hour. This is the time of the golden light, when the sun is still warm but not strong enough to burn. The beach is clothing optional, so I opt for none. I pull my dress over my head and recline on the sarong. Rick sits beside me in his shorts and T-shirt.

Here, in this soft afternoon light, on this magical beach I feel *so beautiful*. Even my movements feel full of grace. I had my baby decades ago, but today I feel like a goddess. In my twenties I may have been more physically beautiful, but I never felt it, never inhabited it in the way I do now. My light brown skin has no tan lines, my hair hangs long and loose, my breasts are small, but they are even and firm, my belly is flat and smooth.

As I write this now, over twenty years later, I can still perfectly recall that day on the beach. I felt young then, as do no longer, but I know that I once had the experience of fully inhabiting a young and lovely body. It is a touchstone.

Just then I could see Sri Juan approaching, with his dark, dark, skin, his black shoulder-length hair, his bare chest, and a white cotton cloth wrapped around his thin hips. We had arranged to meet and talk about the next day's sweat lodge ceremony (called a *temezcal*). As he approached, I could tell that even this young holy teacher felt my power. He squatted with one hand resting on the sand and spoke to me in a respectful manner.

I had met Sri Juan just a few days earlier. I was sitting on the beach after a swim when he noticed me and came toward me. I guessed he was about thirty. "May I help you in any way?" he said. His English was understandable, if delightfully broken. Normally I would wave off someone who approached me like that, but for some reason I was interested.

"Do you want to know anything"? he asked humbly.

Why yes, yes, I did, "I want to understand 2012," I said with a lift of my chin.

Although that date was still years away, I had heard rumors that when the year 2012 came there was going to be some sort of huge shift. Something about a Mayan prophecy. What, exactly, was coming? The age of Aquarius, filled with peace and love? World war? Natural disasters? Should we prepare in any way? I had more questions than answers.

I don't know what would move me to ask a stranger on the beach about 2012, but that is exactly what I did.

"I can teach you about that," he said, as if it wasn't an unusual question at all. He told me his name was Sri Juan. We made a plan to meet later that afternoon, and at the agreed upon time we met back at the beach then hiked up the rocky headland to the overlook. We sat on the ground and my lessons begin.

"What I am about to tell you is just one view of things, and others have interpreted it differently," he began.

I nodded.

Then he used his finger to draw in the dust, similar to Jesus in the parable. First, he drew a dot, "This is one." Next to it he added another dot, "Two. But when you get past four it changes from dots to a single horizontal line, like this." And his finger ran through all the dots. "So how much is this?" he quizzed me, as he drew one line with three dots above it.

"Eight," I said, like a proud grade-school student. Instead of Roman numerals I was learning Mayan numerals.

Juan smoothed out the dust. "Now if I just draw two parallel lines that equals ten, so what would this be?" he asked as he dragged his finger through the dust creating two parallel lines with three dots above them.

"Thirteen."

"And that's as high as we go," he said quietly. "Thirteen days is called a 'wavespell,' with each day in the wave pattern represented by the symbols you just learned. After thirteen you start over again at one. Different points in the wavespell have different energies. Just like the phases of the moon draw out different energies. There are good days for starting things, good days for finishing things, good days for working hard and good days for resting." He looked in my eyes to make sure I was getting it. His eyes were so dark I couldn't tell where the pupils changed to irises. Looking there was like falling into a warm black pool.

"In addition to these numeric symbols," he continued, "each of the days is represented by a different shape, called a glyph." He pulled out something the size of a comic book and showed me a calendar filled with the dots and lines, and the different square-shaped glyphs. They looked like tiles with stylized animals, bubbles, star designs, question marks, and every combination of them. And then in addition to the dots, lines, and shapes, each of the squares was colored, either white, yellow, blue, or red. The result looked like a calendar created by extra-terrestrials – it looked like nothing I had ever seen before.

Juan then asked me about the day and year of my birth. "See how each calendar square is a different color and glyph? Because of your birthdate you are a Yellow Galactic Seed."

I knew I was a Leo, but I never knew I was a Yellow Galactic Seed. "What is that sort of person like?" I asked.

In response, Juan carefully drew my glyph in the dirt. Inside the square was a horizontal line that dipped down in a U shape in the center. Above the U was a small circle. "That dot represents the seed, and the dip is the, umm how do you call it, the place that the seed will be planted in."

"The furrow," I interject, ever the teacher.
"Yes."

"Seeds, like you, like to start things. And the galactic tone means that you harmonize and model integrity."

I wasn't completely sure what that meant, but I was thrilled to be learning this new way of looking at things -- and not from a book or a documentary, but from a warm and patient person that I was feeling very drawn to.

He took out a tangerine, peeled it, divided it in two, and handed half to me. "And there is something else you should know. See how each of the months are twenty-eight days long? This matches with the moon cycle. By counting time this way there are thirteen months in a year. Each month has one full moon and one new moon. It makes sense, not like the calendar we use now that could have two full moons in one month. When you become tuned to the Mayan Calendar you are much more in touch with nature and all of her energies."

"Now, what you asked me, about 2012," he continued, "that is the end of the longest cycle of the Mayan calendar that was started over five thousand years ago. At the end of this cycle everything will be mmm... what is the word?...made larger. Good is stronger but so is bad. There is nothing to fear. Just stay flexible, like those surfers there in the ocean, they can still ride the waves even when they get bigger."

My head was spinning, but at the same time I felt joyous with these discoveries. It was time for us to part, but Sri Juan had become very special to me. I didn't tell him about the warm red light he seemed to ignite in my core, but I did tell him that he reminded me of the Indian holy men I had seen, and I asked about his bloodline. He said he was Mexican, but he felt strangely close to India, and that was why he had taken the title Sri. His real name was not Juan, that was just the name given him by the young men he was traveling with who were offering *temezcals*.

Why was I so drawn to this young man? I was a happily married woman, I wasn't interested in having an affair with him, yet here we were, sharing a special type of love, an energetic love that had nothing to do, with our names, our ages, or our social status. We were bringing the energy of heaven down to earth.

I had been raised in the Christian tradition, but it never really 'took' when I was young. Then, as a teenager, I had a born-again experience and developed a close relationship with the Christ consciousness. But now, years later, that focus on just one holy being had waned, and it was being replaced by a larger, more universal consciousness. I didn't even have a name for it, but I could feel it growing in me, and I could feel it emanating from Sri Juan. My heart was still expanding – it had room for more. Had I met my first guru? I was now in the territory of wild spirituality.

Dear reader, why am I starting the story here? Well, one of the things that makes memoirs so compelling is that the story is not only that of what happened in the past, but also of completing one's picture of the past. Of seeing how it fits in to the life that went forward from there. My picture of this past is a time of personal power. I had a healthy body and a loving partner. I had a PhD and a teaching position. I had a grown child and a black cat. I had a deep, deep love of our planet Earth, and a special connection with the wild plants and animals she brought forth.

Pachamama is the Earth Mother goddess of the indigenous people from this part of the world. When I walk up the beach at sunset, there she is. I see her profile in the rocky headland. When I hike up on the headland, I look down and see her bosom: a huge rock emerging from the sea, shaped like two breasts with a hole in the middle where her heart would be. As a strong wave enters the heart of the rock it creates a booming sound. I call it the 'heartbeat of the world.' All around the rock are white foam trails on the turquoise water. I watch them form and dissolve. I could watch them forever.

The land, the land, the land…it is always here for us, no matter what else happens. Pachamama wears a tiara of cactus, the airy shade of leafleted trees gentles the sun along her paths, ants cross and climb, may it ever be thus.

Burning Man One (age 53)

Rick just turned off the engine, but our adrenaline is still on high. It's three a.m. in the dark desert. The wind is blowing hard and it carries with it sharp dusty matter we hear hitting the van.

After driving through remote stretches of Nevada, hour after hour, in the middle of the night, we finally passed through the Burning Man entrance gate. Then we rolled along slowly past rent-a-trucks, large tents, piles of bicycles, scaffolding, and metal poles with guy wires. Everything that could flap was flapping in the strong wind.

Eventually we find what looks like an empty space large enough for our van. We pull in and cut the engine. Are we safe now? Can we finally sleep? Then suddenly, out of the darkness, a figure covered in rags and carrying a flashlight is walking toward us. The beam is illuminating the white particles flying horizontally through the air. The figure has a black scarf wrapped around its head and covering its nose and mouth. It is heading right for our driver's side window. He (we could tell by now) raps on the window, we roll it down only part way, a bit reluctantly.

"You can't park here, this is *nectar village*," he shouts above the howling wind.

We have no idea what he means (what the heck is a *nectar village*?). But we got the message that we must move. We start the engine, back out, try again.

"There, there's a little spot."

We pull in. Are we OK here?

Apparently not. Five minutes later the same figure appears out of the darkness: "You can't park here, it's *nectar village*," he shouts again. There are many things we didn't understand about where we are, and nectar village is just one of them.

Our next attempt at parking is more successful. This time, minute after minute goes by and no one knocks on our window. The adrenaline starts to subside, we undo our seatbelts and move toward the mattress in the back of the van. Rick, who has done most of the driving on that *dark desert highway*, was ready to sleep, but I have to get a sense of where I am first. I am finally here, but where, exactly, am I?

I slip out of the van's sliding door and start walking. The wind has died down. The sky is clear now, and on the far-away horizon I can see the silhouette of mountains backed by indigo pre-dawn light. I have no idea where I am heading, but on the dirt road I see other shadowy figures passing by, in both directions, on bicycle and on foot. Every seventy yards or so there is an intersection with another dirt road. I realize I could get helplessly lost out here and never find the van again. There is no cell service, so even a "find my car" app won't work. I have to make good choices (as I heard them tell the pre-school kids) and my choice is to walk only in a straight line away from, and then back to, the van.

As I walk, I start to relax and look around. Many of the people I pass are dressed in costumes that include hats, leather boots, and lots of buckles. Their goggles and bandanas are hanging around their necks now that the wind has settled. A few are dressed plainly, like me: hiking boots, jeans, a fleece jacket.

It is mostly quiet this time in the morning. I walk and walk past camp after mysterious camp. Some camps are surrounded by RVs, others have large canvas tents with door flaps tied shut, some are a collection of igloo-type structures covered in reflective foil. After about a third of a mile the camps suddenly stop. The last row of camps are impressively built, with plywood instead of canvas; and they are decorated with colorful paintings and welcoming doorways. But what takes my breath is what I see far across the desert past this last row of camps -- it is The Man! I am at Burning Man and I have stumbled across The Man himself!

By now the sky has gone from black and dark blue to light blue and pale yellow. I feel a surge of energy as I walk the half-mile across open desert to the feet of The Man. It is a beautiful cool, calm, morning in the desert. Even this seemingly 'blank' part of the Earth carries with it a powerful energy. I reach the huge wooden sculpture with the triangular head just as the sun rises over the horizon. The sun is orange, but golden rays stream up from it into the clear jay-blue sky.

What magic! Could I have planned to be at The Man on the very first morning of Burning Man, just as the sun rose? Well maybe, but it would have taken many hours of intense planning and stress, striving to meet that goal, and very possibly it wouldn't have come to fruition anyway – because plans are like that. Instead, I have literally stumbled my way here. *Hallelujah*!

A few other characters are scattered around the base of The Man as well, but it is a sparse little group. Nothing like the hordes of thousands that would gather here later in the week. How many of these others had planned to be here this moment? And how many have just had it happen to them, like me? It is impossible to tell, but here we are, temporarily bound in a community of awe and appreciation. It hasn't taken long for me to discover the draw of this place. Mystery and magic are everywhere.

When I finally turn to head back, I carefully navigate toward the road I emerged from. But now, with the sun shining, everything looks different. At the corner of "my" road is an open-air roller rink. Wait ...*a roller rink*? Yes, complete with skates to loan. A pair of women, wearing very few clothes, are testing it out; one, in a fur bikini, has long blonde braids that are streaming behind her, another is in an aqua-colored dancer's tutu with a pink tube top. I try not to stare. As much as I enjoy skating, I am not about to join in. But, wait a minute, *what*? Someone decided to build a temporary roller-rink in the middle of the desert? Of all the things.....whose idea was *that*?

It was free of course; everything is free. The rink builders had done this at their own cost, just because they wanted to. And if that were not strange enough, next door was an open-air bar with barstools that had seats six feet off the ground, and a bar seven feet high. What the, *what*? Odd sights were everywhere as I made my way back down our road. I was passed by a small vehicle shaped like a snail, and then another one that looked like a cartoon-duck on its back, with orange feet sticking up in front. This was no haphazard construction. Someone invested much time and money into creating a cartoon-duck vehicle. Who would do that??

Eventually I realized I was surrounded by people who would do that, but more about that later. I have not slept at all and it is time to make my way back to the van. I doubt that Rick is awake, but if he is, he will surely be wondering where I am. (Where *am I?*) When I reach our white van and slide the door open, I see that he is sleeping comfortably on the mattress in the back. But I still have wild energy moving through my body. I undress and crawl in next to him. As he wakes, I start telling him about my adventure, but soon my lips are telling the stories to his skin instead. As always, our bodies are good at listening, and responding, and telling the deepest and best stories to each other. Eros has her way until I, too, finally drift to sleep.

Later I would learn more about the current of erotic energy present in this wild desert. But I'm glad I had no idea about it before I discovered it for myself.

I had been reading about Burning Man for years. It's difficult to imagine now, but Burning Man was barely mentioned in the commercial media when I first learned of the wildly creative and lawless gathering in the Nevada desert. Part of me was put-off: why would I want to camp where there was no water, no plants, basically nothing living? Being the nature lover that I am, when I camp I love the feeling of being supported and protected by nature: the moss beneath my tent, the stream for collecting water, the twigs to start the fire, the birds singing in the morning. There would be none of that at Burning Man, only strong sun and white powdery dirt. But another part of me was intrigued at what our human culture had created. The event seemed to be the cutting edge of our culture and I wanted to see that and experience it for myself -- the wildness of our culture.

By the time we made it to Burning Man we'd been married for thirty-one years. 'Happily married' sounds like a cliché, but in our case it was true. Happy, and happy to be married.

Due to my dogged independence, and perhaps my fear of total entwinement, we had kept our finances and our work lives separated. This was functioning wonderfully -- we never argued about money or staff. But recently, after all those years of keeping boundaries, I felt ready to embark on a project together. We applied for, and were awarded, a two-week residency at a cabin in Oregon. It was for two people who wanted to work together on a creative project. Rick was a photographer and we planned to use our time at the residency to create a small lyrical book about Western old-growth forests. This late-summer journey to the Western old-growth would also lead into my sabbatical (finally!).

There must have been a few other people in my community in the late 90's and early 2000s who wanted to visit Burning Man, but I didn't know a single one of them. In those days, no one I knew had even heard of it. For an academic in the east, Burning Man falls at the most impossible time -- the week leading up to Labor Day, the week of drop-add schedule changes, the week of finalizing syllabi, the week of important faculty meetings. I could only dream about attending. But with a fall sabbatical and a residency in the West, maybe, just maybe, 2009 could be the year I experienced Burning Man.

When we reserved our vehicle for the month out west, we rented a Chevy cargo van just in case we had an opportunity to go. I had no idea how to get tickets, and didn't even bother trying, but regardless of if we made it there or not, the van would be convenient for camping, it had two seats in the front and was completely empty in the back.

Long before glamping was a word Rick and I had devised a way to travel and enjoy wild places while still being comfortable. The key was to fly from east to west with a big duffel bag. Inside the duffel we'd pack two sleeping bags, a small tent, a screw on burner for a propane stove, headlamps, water filter, and minimal kitchen supplies (a cooking pan, a few plastic dishes, and utensils). With that camp-in-a-bag we were ready to go anywhere. The duffel was rather heavy and awkward, but at the arrival gate of the airport, one of us would stay with it on the luggage cart while the other would go pick up the rental car and pull up next to the curb. The camping bag went into the rental vehicle's trunk, and we were set for any adventure. We did this numerous times all throughout the West, flying into Salt Lake City or Vegas or Denver or Bozeman or Seattle or...

you get the idea. We were an adventurous couple. Rick loved to visit new places because they stimulated his photography. I loved the adventure, but mostly I loved the nature. I loved hiking to hot springs with the smell of desert sage in the air and the sound of chittering grasshoppers in my ears. I thrilled seeing a new-to-me wildflower. I loved lying flat on my back and appreciating the splay of ponderosa pine needles across the moon. I loved seeing herds of antelopes, and, of course, I loved being in the old-growth forests filled with huge trees and intoxicating air.

This time, our first stop was Seattle, to spend a few days participating in a wedding.

I had been watching Craigslist, a popular website at that time for buying and selling, and I noticed that every day a mattress was being given away. My plan was to grab the mattress being given away the day we were leaving Seattle and put it in the back of the van. I was so sure of this plan that I had packed sheets and pillowcases in the duffel.

But the morning we checked out of our hotel there were no mattresses listed. Bummer. I was thinking my plan had failed, but I was still scanning Craigslist on my phone as we drove away. Suddenly a listing popped up and I texted right away that we wanted it.

That mattress could have been in any condition, and we'd have taken it, but it turned out to be a clean, new, double mattress that fit perfectly in the back of the van. On went the sheets, out came the sleeping bags, we stopped at Walmart for cheap bed pillows and a small propane cylinder, and we were good to go.

So, that is the couple who flew to Seattle, put their camp-in-a-bag in the back of the Chevy cargo van, got a used mattress from Craigslist, and headed toward a creative residency focused on old-growth forests, with a vague dream of experiencing Burning Man (for one of us anyway). But we were not kids, Rick was sixty-seven and I was fifty-three.

When we arrived at the residency cabin, I was happy to discover that it was more like a chalet than a rustic cabin. Our two weeks of hiking the forested trails, taking photographs, writing, and making side trips to old-growth forests, went blissfully. By the end of our time there we had a mini book assembled. We worked well together. Whatever was I afraid of all those years?

Big leaf maple, I see you, I am called to touch you. You with your mossy green stems splaying wide overhead. In your green-lit company I feel energy, and messages without translation. I am not alone. You are not alone. Together we will face this world.

I haven't really described Rick yet – with his short curly black hair, green eyes, and white-toothed smile -- he is devastatingly handsome, in that swarthy *most-interesting-man-in-the-world* way. He is also an extrovert. He loves people and they love him. I'm the opposite -- an introvert happy in solitude or with just one other person – so, although we were having an excellent and creative residency together, by the time someone new knocked on the door Rick was ready for some novel social input.

There at the door stood Don, the caretaker, also with curly dark hair and a glowing white smile. Don was interested in us and we were interested in him. The energy of the conversation kept growing and growing. I told him I was writing about old-growth forests, and he told us he was getting married in an ancient redwood forest in a week (!). After the wedding they were going to honeymoon with family and friends at Burning Man (!). I was getting dizzy at all the connections.

Rick did lots of wedding photography, and he asked who was shooting their wedding. They didn't have a photographer.

I mean, come on, what are the odds of all this? By this time the vibe was so high it felt a bit like falling in love. Rick offered to do the photography in trade for two tickets to Burning Man, and without even asking the bride-to-be Don made the deal. See you next week Dear Don. And that's how it happened; we were going to Burning Man.

Reader, does it work this way for you, too? Dot to dot, a thread, a dream, that then becomes reality? Is it because of something you have done? Something you are? Or are their larger forces at play and you are just a pawn? Are we all just riding a wavespell? I notice the tenses in my writing going from present to past to present again. It feels like a giant mobius strip that keeps feeding back onto itself.

When we finally wake up in the back of that van, in the middle of a Nevada desert, it is hot. The sun that had risen so majestically over the colossal three-dimensional hangman is now pounding straight down on our metal van. We want to venture out, but another storm is blowing white dirt everywhere. They call them 'dust storms' but it's more like dried clay, salt, and tiny sharp rocks, all moving horizontally through the air at thirty-five miles per hour. It gets into every orifice there is, dead or alive.

Rick wraps up his camera equipment in cloth sacks and plastic bags, then stores it deep in the van where it will stay, hopefully dust free, for the remainder of our days here. Listening to the wind-carried particles scraping against the van's paint, I am glad we are in a rental vehicle, and glad that the Seattle rental agency has no idea where we are.

We move back up to the front seats to get a better view, and as we watch small tents being blown away, we are grateful for our steel home. The few people we see walking by look like they are from an alien planet. They wear goggles over their eyes, masks over nose and mouth, and scarfs wrapped around their head and ears (yes, the dust even goes there). Most are wearing small hydration backpacks filled with water and they occasionally suck on the clear plastic tubing. There is not a single plant, or stick, or bird, or insect anywhere. This vast, flat, open space has been nicknamed the *playa*. This is nothing like the camping I had been used to, where I hung my food bag in a tree to keep it from the animals. We are not prepared with any of the equipment that seems necessary to survive here, we are clueless newbs.

Our days between the writer's retreat and the wedding were spent in Eugene visiting friends. When we told them of our plans, they directed us to their across-the-street neighbors who went to Burning Man every year. Burners! How cool to be out west where people had not only heard of it, but they were actually attending. The neighbors were super friendly. Lydia showed me the jacket she had just finished making for her husband. Into the back of a nice wool suit jacket (for those cold evenings) she had hand quilted layers of hearts, each diminishing in size and outlined by red glo-wire. The lights were important so he could be seen in the darkness – there are no streetlights there. The hearts also had a deeper significance. She dove into a story about their previous year's trip...

"Burning Man can be hard on marriages," she stated flatly.

One morning at Burning Man her husband returned at four a.m. sheepish and sad – he had lost his wedding band.

"Where were you when you lost it?" Lydia asked him.

He had been at Thunderdome, a camp installed at Burning Man every year since 1999. It is a large, Buckminster Fuller type geodesic dome made of steel bars. Hanging from the steel bars are heavy-duty bungee cords with harnesses attached. Volunteer participants are strapped into a harness, given a giant foam bat, and flung out into the center of the dome to face another suspended and 'armed' participant. Meanwhile onlookers climb up the outside of the dome and cheer them both on. It is loud, high-energy, and not for someone afraid of getting hurt. The later in the night it gets, the more otherworldly the scene and the screams. Even walking past it is an experience.

He guessed that his ring had fallen out of his pocket while he was a combatant in the dome.

As soon as there was light in the sky Lydia headed to the dome, which was now empty and quiet. She started on one side, methodically dragging her foot in a line across the powdery ground surface inside the dome. Line by line her search area expanded. At last, during one sweep, a ring was revealed in the dust. She reached for it, and was happy to have found it, but a few questions had formed in her mind during the time of the search. The biggest was: "What was your ring doing in your pocket instead of on your finger?" Men slip their wedding bands into their pockets for one main reason: they want women to think they're single. When Lydia got back to their camp she had a question for her husband: "Do you want to be married, or not?" Burning Man is not all fun and games.

I'm thinking back to her story while sitting in the front of the van, watching the dust storm, and feeling the temperature rising higher and higher. The unclouded August sun is shining in the windows that we dare not open lest we have a bed filled with micro-gravel. We put up sun shields where we can – sarongs or pieces of cardboard – and sit...and sit, for hours. Even with the fascination of where we are, and what is going on around us, it is getting uncomfortable and boring.

Are we just going to sit here and bake?

Rick's knee had taken a turn for the worse in Eugene and he couldn't walk far. Finally, I can take it no more and I venture outside of the van. My goal is the camp next door. We have been looking at the back of their large dark green plastic tent for hours. I turn the corner and see that a flap has been propped open. "Come in, come in," a handsome man in his fifties urges. "What would you like to drink? We have a full bar."

I am stunned. The spacious room is shady and cool, and relatively dust-free. Around the edge of the tent are beds with *satin sheets* in case anyone wants a rest. A fan is wafting cool air. Crosby, Stills, and Nash are playing on the high-quality sound system. The three men behind the bar are kind, and generous. I have fallen through that magical looking glass again. *We were baking in a van while this was next door!?* One of them makes me a rum punch *with ice*. Free of course, everything is free. Once you are here no money changes hands.

"We're all doctors," my bartender says. "We do this once a year just for a change of pace."

Even more thrilling than discovering what was next door is my excitement at being able to blow Rick's mind. I excuse myself to walk back to the van. "Come on," I say, "Get out. Don't ask any questions. Just follow me."

Although we didn't know it then, we eventually learned that there were hundreds of bar camps like this, each one with its own unique theme, specialty drinks, and musical backgrounds. But *this* bar camp, our first, will always be the most special. My second favorite playa bar experience was in 2011. But that story will have to wait. By the time we finished some cold drinks and some deep conversation with the doctors the dust storm had stopped, and we headed back to the van and rolled down the windows.

A Ryder rent-a-truck pulled into the vacant lot across from us. (The *Ryder* name blocked off with electrical tape since advertising of any kind here is considered uncool.) Then an RV pulled in, then another, then a 250-gallon water tank got delivered and filled by a Nevada company based 100 miles away. This was rather interesting. For a long time we just watched through our big windshield, but finally we decided we should meet our neighbors.

They were about the same age as us– older than the majority of Burning Man participants. Older, *but cool, like us*, we think. James, the dad, is in the business of patenting and manufacturing medical devices. When those devices hit, they hit big. And James's had hit big a number of times. His pampered daughter, Stephanie, and her new husband (in the second RV) had just celebrated a big wedding.

When it came time to make honeymoon plans Stephanie blinked her blue eyes at her father. "Well… Daddy," she said. ("Oh boy, here it comes," he thought, "she wants me to foot the bill for a trip to Tahiti.") But the ask was even bigger than that, she wanted something that would bond the whole family together AND share their largess with others. She wanted them to create a special camp at Burning Man.

Daddy said yes, although he had never been to Burning Man, and had never had a desire to go. The family brainstormed and came up with their gift to the burner community: in a place that was frequently uncomfortable, hot, and dry, they decided to create a cool, comfortable oasis.

The following morning the truck with the big tent arrived, after it was set up, the rugs arrived, then the guys who spent all day hooking up sprayers and twinkle lights in the ceiling of the tent (run from a power pack charged by a generator). Finally, the furniture arrived – about ten white bean bag couches. This was blowing my mind. All of the stuff was *brand new;* brand new, beautiful rugs that would be completely ruined by the end of the week. But that was the plan, that was the gift. Entering this lovely space in the heat of the day you could recline on a couch and be cooled by a constant vapor mist that evaporated into the air.

This generous family soon became our friends. We would bring food over to their grill and dine together. As the days went on James laughed about what they thought would happen in their created space, and what actually happened there. They thought people would stop in for a quick rest, get refreshed, and move on; but one petite Japanese girl came in, stretched out on a couch, fell asleep, and was still there eleven hours later. They never considered kicking her out. They were delighted that their place was such a respite for someone experiencing jet lag, culture shock, and desert air for the first time – and who knows what else.

As difficult as it was for Rick to walk, it was equally difficult for us to stay put during this once in a lifetime opportunity. One night he limped far out in the desert to look at the art pieces when suddenly his knee began hurting terribly. We paused in the middle of a wide-open area to discuss what we should do -- how to get back across the vast distance considering the pain he was in. I suggested that we needed to find one of the green 'community bicycles.' And before the words were out of my mouth, we looked down and there was a green bike at our feet! We were both incredulous – but we both reacted differently. My instinct was to not question the gift, but to just get on it and go, before it disappeared as quickly as it had appeared, but Rick tried to understand: "Was the bike there when we stopped walking?"
"No."
"Did you see anyone walk up near us?"
"No."
"Really?"
"Really."
"*Really?*"
"Really." But reality was starting to feel slippery.

The world is a magical place. I hope you know this. You don't need me to tell you. In the end we both thought of the bike incident as a miracle; like

Jesus manifesting wine, loaves, and fishes. That's what it felt like. Or was it a different type of miracle: a miracle that our minds were so attuned that we saw things exactly alike? Burning Man was not hard on our marriage; indeed, it entwined us even more deeply together.

Before he could get on the bike, a tiny electric cart designed to look like a butterfly with big, colorful, wings, pulled up next to us. "Would you like a ride?"

Yes, indeed, we would. We climbed onto the bench seat just big enough for two and were transported across the wide desert on the back of a butterfly. On the way we passed the bar with the six-foot barstools. It was called the Stilt Bar and it was made for those who were wearing stilts. What the, *what*? The bar was filled with stilt walkers having a great time. We were delivered right to our van, for free of course.

It was kinda cool to be adrift in Burning Man, just the two of us old virgins (that's what they call first-timers). Even though we weren't zipping across the playa, like many of the other attendees, there was plenty to see right in our 'neighborhood.' One day we saw a pack of about fifty people running by, men and women, completely naked from head to toe and covered in gold paint. What the...? We made friends with folks in another neighboring camp who made pickled eggs all year, and then gave them away at Burning Man (one of them was the guy in the chicken car).

One morning another large rental truck pulled up nearby. The people who gathered around it were not dressed like the other attendees in colorful wigs, hair ribbons, and fish-net stockings; this group had on steel-toed shoes, leather gloves, and safety glasses. They were serious. Out of the truck came long hollow pipes, huge gears, and heavy steel beams. There was banging and grunting and directions were being shouted by the guy who was obviously in charge.

I was fascinated by all this, although too shy to interrupt and ask about what was happening. I eventually learned that they had driven all the way from Vermont to assemble a gigantic human-powered Ferris-wheel-like thing that would roll across the desert with participants looping around in the huge steel wheel. *What?* Who does that with their time and money, their life? For sure the designer spent all year fabricating this thing and spent untold dollars to build it, not to mention getting the whole thing and the construction crew to the other side of the continent. Did he have a wife and kids? If so, how many times did he choose this Mad-Max Ferris wheel over them?

More than anything, Burning Man opened my eyes to this possibility in life: shit doesn't have to make sense. When I got home and people asked me what it was like, all I could say was that it was like going to another planet where the currency was creativity instead of money. At Burning Man I was nothing compared to wheel guy, or the chicken man, but in the 'default world' (as burners call everything outside of the entrance gates) our social status might be reversed.

One day I discovered that around another corner was Nectar Village. *The* Nectar Village. Turns out that the mysterious Nectar Village was a collective of smaller groups that all organized to make a big mega camp together. It had an astonishing fifty-foot-tall metal sculpture of a spirit woman at the entrance. Her 'hair' was created from thick metal chains. Dozens of bikes were parked at her base, they belonged to the participants who were inside where there were camps for yoga, acro-yoga, chanting, healing, finding your power animal, and a steam bath.

After living in a hot van in the middle of dust storms, I was getting quite ripe. I happily got in line at the Steam Bath Project. Eight people at a time were given a space to strip down and stash their clothes. Off came boots, and everything else. We were then led into a small dimly lit dome. Inside it was steamy and there was a hose of warm water that we passed around to wet our entire bodies. It felt so, so, good to soak my scalp and my long, dusty hair. Next, we passed around the lavender soap, and then the hose again for a final rinse. When we were all done, and I stepped outside, I felt renewed.

For the rest of my time at Burning Man, Nectar Village became my home base. Many parts of Burning Man felt dark or frightening to me, but Nectar Village was cleanliness, goodness, positivity, and light. This was an ecosystem I was in, but one made entirely by humans.

The bride and groom from the redwoods had things to deal with at home after the wedding, so they arrived a few days after us. They had bikes and all the rest of the usual props – having been to Burning Man many times before – but one evening they planned to spend their precious time with us, and they generously went on foot with this bikeless pair. A friend of theirs offered us some MDMA – the pure active ingredient of Ecstasy tablets. We each took a small dose and then wandered into the desert at the speed of Rick's limp.

I haven't described the lights yet, the beautiful, colorful, lights in the dark, flat, desert at night. Things are lit for both beauty and safety -- unlit bikes and people can, and do, crash into one another. There are lights of every color on bikes and clothing, on art cars, and sculpture; and sometimes the art is made completely of colored lights. It is all at once delightful and disorienting. *Which direction is camp, again?* As we walked slowly through this sea of black punctuated with color we encountered an art piece.

There are multitudes of art pieces at Burning Man, and they are spread across such a wide area, that no one has ever seen them all during the week of the event. And after the event is over every single one is removed or burned down. Most are never shown again. A few are in private collections. In 2009 hardly any were acquired by museums or galleries, but that has changed. We could go deep into the rabbit hole of the art there, but I'll just say it is my favorite part of the experience, and the art has a mysterious life of its own.

The art piece that the four of us encountered was a huge made-of-steel flower.

As a botanist I know that many plants have ovaries in the base of their flowers, where the seeds form. Flowers can have ovaries with one chamber, or two, or three, or four, or eight, etc. Flowers vary a great deal. But this magical, giant flower has its ovary opened into four chambers, and each chamber is lined with a soft red fabric, just like the lining of a womb, and each chamber is just the right size for a human to back up into. Without saying a word we each back into our own soft red chamber. Overhead, at the top of the sculpture, a steel butterfly is emerging from its chrysalis. Here, at eye-level, we stare into each other's eyes. Newly-married, long-time married, newly friends. I see you. Eyes to eyes to eyes to eyes, we are deeply in love with each other, with ourselves, with a world that can manifest all of this.

That memory is another touchstone. By the time I returned to Burning Man, two years later, Rick was dead.

Burning Man has a different theme every year. In 2009 the theme was Evolution. It is interesting that I experienced our "loving eyes" sculpture as a flower, but when I recently did some research on it, I learned that it was supposed to be a human womb, and not the ovules of a plant. Perhaps the red lining should have clued me to the fact that this was mammalian reproduction being represented, not botanical. They are both analogous, of course, as reproductive spaces where evolution becomes manifest.

Now I can look online at images of Bryan Tedrick's sculpture, *Portal of Evolution,* and see the stylized human ovaries thirty feet above the 'womb' and the flower-like fallopian tubes another ten feet higher. A butterfly on the top, speaks to how each manifest creature eventually grows its spirit wings and flies on. We didn't know that night would be the last time the four of us would be together before Rick's spirit would fly on. We didn't know that Don and his new bride would spend many years, and many dollars, trying to conceive. My love for all three of the humans who shared that womb with me remains as real today as it was that night many years ago.

The Human Ecosystem

In my first book, *Teaching the Trees*, I wrote about all the connections between plants and animals in the forest. I explained that without holly trees there would be no holly leaf miner, and without leaf miners there would be no parasites of the leaf miner. Both of these insects, and the holly berries, are food for birds. Likewise, without beech trees there would be no beechdrop plants, and without those plants, one very tiny gnat species would be without a food source. For me, it was easy to see the connections all through the forest. But what about in the human ecosystem? Could I understand our connections in the same way?

"Were you close to her?" the woman across the table asked, when I told her my stepmother had died that morning. We were sitting at the communal table in one of my favorite Zipolite cafes.

Surrounding us was greenery, hibiscus flowers, and the sounds of birds and waves.

Was I close to her? Yesterday I might have answered differently. The woman across the table - - a new acquaintance - - was just trying to make socially appropriate conversation, but the scientist in me was trained for truthfulness - - even if it meant admitting uncertainty.

My stepmother had been a major force in my life for forty years, and her thoughtfulness was boundless. Even with a clutch of seven children to consider (and that was before the nine grandchildren came along) she never missed a birthday or a holiday, we even got Thanksgiving and Halloween cards. (I can't make the same claims about my consideration for others.)

My stepmother thought of others a lot, too much, I would have said in my adolescent years, when her attention was so focused I thought it was downright creepy. Taking things at face value was something she did not do. In her world every glib comment or forgotten occasion had a deeper significance. She would read more into my actions than I had intended in their commission - - a trait that irritated me *majorly,* as I would have put it then.

But this woman loved my father with an undying and selfless love. She was there to celebrate my marriage and the birth of my baby. On her vacations she bought cute little outfits for my daughter. And the sweaters! She knitted baby sweaters, sweaters for me (four), and my husband, and my brothers, and my sisters (don't forget there were seven of us), and their spouses, and their children. I have never knitted a thing in my life.

She was a special woman - - but, still - - there is something about a blood mother that cannot be replaced. Is it those first formative years? Or the genetic bond? Whatever it is, long years of selfless devotion cannot equal it.

The whole family had gathered for Thanksgiving only a few weeks previously. At that time my stepmother was very ill, but still getting up and dressed every day. She sat at the table with us, eating turkey and drinking champagne. A week or two later she was spending her days in bed. When I spoke with my father on the telephone he said it was a temporary setback - - something to do with her blood sugar. I thought differently but didn't have the heart to tell him. I suspected she was tired of fighting, as I would be, as anyone would be.

The diagnosis of lung cancer after fifty years of smoking had been bad news, but not altogether unexpected. See how heartless I am? Even seeing her so ill at Thanksgiving and hugging her - - knowing that it may be the last time we would be together - - I did not cry even then. When I left a few weeks later for our long-planned trip to Mexico, I carried with me the possibility that she wouldn't last until I got back. I went anyway.

During the first week in Mexico she crossed my mind occasionally - - and of course there was sorrow for what she and my father were going through - - but I was certainly not wracked with guilt while swimming with schools of fish in the greenish-blue sea, or while lying in the hot sun, or while sipping my cold *cerveza*, or while drifting off to sleep to the sound of waves.

Then one night, in my mosquito-net covered bed, I dreamed that I was at a small gathering of people. We were all saying good-bye to one another in preparation for leaving. And there she was.... "Of course!" I thought, "I must say good-bye to you." It was as if we both had that thought at the same moment. We approached each other knowing that this was a very special good-bye - - the last we would ever say to each other. The *big* good-bye. We embraced and soon we were both crying.

"There was a long story I wanted to tell you," she said, "but there's no time for that now. There's only time for good-bye."

In that dream space we were holding each other, loving each other, crying softly. Suddenly a strong charge of energy, like an electric rod, connected us from heart to heart. I had never felt anything so physically unusual in a dream. The charge was so strong it woke me up; and when I woke in the darkness, in that bed in paradise, thousands of miles from my stepmother, I was sure it was a real good-bye. Only then did I cry and mourn my loss, our loss.

When the sun rose, I walked into town to make an international call. My brother answered. "I'm calling from Mexico," I said. "Is mom still alive?" I was surprised when he said yes. "But it looks close," he admitted. Next I spoke with my sister who was also sitting bedside. I asked her to share my dream with my stepmother - - if there were any moments of lucidity left. Then, right there in that Mexican internet café, I began sobbing so hard I couldn't speak. Grieving, I hung up the phone. My stepmother died a few hours later.

I feel certain that we shared a telepathic moment during her last night. Maybe when she knew she had reached the end she spent a few moments loving each of us and wishing us well. That would be like her, wouldn't it? Or perhaps the night of my dream, the moment of our embrace, was when her soul departed - - leaving her physical body behind for a few more hours. As my father would say later, *there is an indefinable time between when someone actually dies and when they stop breathing*. I have no doubt that her soul and mine were in communion, but my scientist self continues to wonder about the mechanisms, about the *how*.

Sigmund Freud said that "sleep creates favorable conditions for telepathy."[2] Our minds are most relaxed and receptive then, undistracted by sensory input. Other researchers, attempting to quantify telepathic experiences, found that, "of over 7000 spontaneous cases of ESP studied in the United States, nearly two-thirds of the cases were dreams...."[3] They also noted that emotional bonds are a predictor: you are much more likely to have a telepathic dream about someone you are close to, someone you care about deeply.

So, to answer your question, lady, I'd say that we were very close. And perhaps that dream message was her last gift to me.

In this story I see my coolness, my aloofness. (How do I pronounce my last name? Like aloof with an M.) But I also see my growing recognition that we are more connected than we realize. We are parts of the same ecosystem, sharing the same planet, and energetically connected by the same consciousness. As the years roll on this lesson returns to me again and again.

Losing Rick

We had been going to Zipolite for ten years now. Our fidelity to one vacation place was something we never had before, but it was something marvelous. We *knew* the paths, we *knew* the birds, we *knew* the waves, and soon we knew the people – more of them every year – a number of them we eventually considered close friends. Some of our friends lived other places, like us, and just came for long visits every winter, while other friends, both Mexicans and non-Mexicans, had made Zipolite their home.

In the fall of 2010 Rick and I purchased tickets to arrive in Zipolite on December 27th, but that was the exact day he died. He was sixty-eight and I was fifty-four. Neither one of us went to Mexico that year. For a while I thought I'd never go back again. How could I go without Rick? It was our special place.

I met Rick when I was nineteen and he was thirty-three. We moved in together when I was twenty-one, got married when I was twenty-two, and had a baby when I was twenty-three. We were married for 32 years. If I could distill my whole relationship with him into a single moment, it would be one in an unknown year, later in our marriage, when I was in bed and turning off the light to go to sleep. "Good night, I love you." I said, as I had said so often, and as he had said so often. But in that particular moment I was aware of how deeply I meant it. How glad I was to have him beside me. How right it was that he was my partner and we were sharing this life together. Then I pulled the comforter over my shoulders, smelling our body scents mingle as they had for decades.

Love can be that simple, and that fulfilling.

If you die from a disease there is always a first symptom, only you don't know at the time that it is the symptom that will lead to everything else. Rick's first symptom was a pain in his rib. My amateur diagnosis was pleurisy (an inflammation of the membrane between the lungs and the rib cage). When he finally went to the doctor they did X-rays, and diagnosed a pulled muscle, for which he was prescribed a muscle relaxant. Both the doctor and I were wrong, but we didn't find out for a few months just how wrong we were.

The pain didn't improve. Like many annoyances, Rick just learned to deal with it. His knee was another one of those annoyances. Pain and cortisone shots were a part of his life.

In addition to the rib pain, a few months later the knee pain progressed up his leg. One day it hurt so much he asked me to get the crutches down from the attic, and he made another appointment with the orthopedist who had suggested a knee replacement.

He was dead three months later.

The cancer had started in his kidney, who knows when, with no symptoms whatsoever, then it showed itself in his ribs, but we missed it, and then in his femur. The pain in his leg, that he thought was knee-related, was a tumor in his femur. When the orthopedist did an x-ray they finally saw it. By that time his leg was about to break, and the follow-up scans showed cancer all over his body. It was stage 4, and as Rick remarked, "there is no stage 5." But he was an officer and a gentleman, and he never whined.

This is his final journey, and however he wants to play it is how he gets to have it, I resolved.

At our first visit with an oncologist the young doctor came in shaking. She knew she was looking at a dead man walking, but she did her duty and outlined a plan of treatment that involved strong drugs and weeks in isolation in an intensive care unit. I asked what the odds of success were and she guessed ten percent. Then I asked if she had ever, personally, seen anyone cured of such a condition and she lowered her eyes, "no."

After we left her office we quickly decided *fuck that*, if he was going to die it wasn't going to be in an ICU unit plugged into machines under fluorescent lights.

Everything happened pretty quickly then. We went to the National Institute of Health in Washington DC to see if he was a candidate for an experimental treatment. When he got there they checked him in because his leg was now so delicate it could break at any time – and it did break when they tried to transfer him to a stretcher for an MRI.

The experimental treatment plan was cancelled due to lack of funding, but we stayed in DC for the surgery to repair his leg. When that was over, he was released to a rehab facility, and then came home a week later, in a wheelchair, for our annual Halloween party. (Why wreck a twenty-five-year long tradition for a little thing like dying of cancer?)

A few days later we flew to Tijuana, Mexico, to an alternative cancer care clinic. The clinic had been recommended by a friend who went there and had his prostate cancer go into remission. I don't know if either one of us was thinking it would do any good, but it felt a bit better than "doing nothing."

In the clinic we had our own comfortable space where we could share a bed, and all our needs were taken care of -- vastly better than running back and forth to a hospital. We stayed for five weeks, and despite all the treatments they were giving him, I could feel him slipping away.

I had taken a leave from the university, and I was no longer wishing time away. I was now clinging to every moment we had left. His body was changing drastically, but my body was changing too. For the previous year I had woken in the middle of every night sweaty and hot. I simply threw back the covers and recognized my symptoms as menopause. I was nearing the end of the fertile phase in my life. At the Mexican clinic I had one last menstrual cycle. After he passed I never bled again.

I walked around the dirty, crowded, border town every day for an hour, just to get some exercise and to avoid going crazy. Along my route was a fabric store I routinely stopped in. I don't sew, and didn't plan to start, but feasting my eyes on the colors and patterns felt nourishing. During those daily visits I fell in love with a Chinese brocade that had a pattern of multi-colored dragonflies on a turquoise-blue background. There was just something about it that delighted me. Day after day I stopped to admire it. One day, I decided that I should buy it to take home; though I had no idea what I would ever do with it.

Since I was going to make a purchase anyway, my mind jumped to thoughts of Rick's shroud. I planned to lay him out at home when the time came, and this would be my best opportunity to get the cloth I needed for that.

I returned to the clinic with all the dragonfly fabric they had left on the bolt -- about two yards -- and seven yards of beautiful, perfect, woven white linen. I showed Rick the dragonfly fabric, but I hid the white in my suitcase.

Even as separated from nature as I was in the city, even here, Gaia speaks to me through her images. Dragonfly, you lovely creature, for hundreds of millions of years you have called Earth your own. You are older than the forests, far, far older than humans. What messages are you bringing to me?

The week before Christmas we flew home. It was a rough trip for Rick, he must have been in pain he didn't want to talk about. Our home was warm and welcoming when we got there at last. At my request, our friends had set up a beautiful space with a hospital bed in the living room. He was so relieved to be there! I knew he was very ill, but I had no idea he'd be dead in a week.

The day after we got home I panicked – I had no clue how to care for a critically ill patient. I needed nursing help but it was Christmas week and it looked like immediate help was not possible. I called a nurse friend from out of state, and he told me that it sounded like it was time for hospice. I told Rick about the conversation and he calmly and quietly said, "hand me the phone."

He had friends at hospice because he had done photography work for them. The number was in his phone already. He called to tell them it was his turn, and the woman he spoke with broke down in tears. That was December 21st, the shortest, darkest, day of the year.

Soon I was learning about bed pads and morphine. Family and friends came to visit, but Rick had stopped eating. I didn't say anything to him, or anyone else, about that, but when he stopped drinking water I knew his time was close.

Christmas Eve, when our daughter and I opened a bottle of champagne in front of the twinkle lights, he finally admitted he was dying with the comment: "I love you both so much, if you gotta go, this is a pretty good way to go." And that was the extent of our conversation about death, once we knew it was on our doorstep. We had discussed it so many times in the years before that I felt no need to press it now, and obviously Rick didn't either. This was his rodeo, I reminded myself, and the best thing I could do for him was to just ride along.

Was I in grief? Yes, certainly, but it did no good to express it. I wasn't one of those who hovered over the bed saying, "don't leave me, don't leave me." I knew he was leaving this world in the same way that I knew, when I was nine months pregnant, that I was going to have to go through labor and delivery.

Accept the inevitable, I told myself. The trees do not mourn the leaves that drop from their branches in the fall.

We often read of people who die, *surrounded by loved ones,* but I didn't want that. I wanted it to just be he and I.

Have you experienced being able to read the thoughts of someone you were in relationship with? Maybe the same question comes out at the same time, or a random topic is being considered by you both. This was a very common experience for us, and it was to happen one last time.

Two days after Christmas there was a lull in the house. Our daughter had gone to the pharmacy to get liquid medication to replace the pills he could no longer swallow. A few other friends and family members were on their way and would arrive soon. In light of my desire for a private death I *thought* to myself: *Well, if you have to go, this would be a good time.* And immediately Rick said out loud, "OK, let's get this over with." He was completely conscious and aware. I didn't have to ask what he was talking about. I pulled my chair closer to his bedside.

This was <u>it</u>. It was time for last words. My brain started searching, what does one say?

All I could come up with was, "You are so loved. You will be greatly missed." But he raised one finger, as in the koan, and I knew I was on the wrong track. This was not a time for him to focus on <u>this</u> life, that he was leaving behind, it was a time for him to focus on his journey to the <u>next</u> life, with no regrets about leaving this one behind. My mind shifted gears. *You are going on an amazing journey,* I thought, *and I am going with you as far as I can.* I had no doubt that he could read my thoughts. He took three more breaths and that was it. Quietly he slid into the next dimension, and I was with him. The room took on a golden translucent light. It was completely still, peaceful, and calm. Outside the window snow swirled in mini-tornadoes across landscape.

Later I learned that Paiute Indians believe the spiraling plumes of dust across the desert are the spirits of those who have departed.

There in the yard, the big oak tree is still standing. It is still alive although Rick is not. The trees are outliving us. This is how it has always been.

I'm not sure how long we stayed in that state of grace, but eventually I heard a knock at the door. It was our first visit from a hospice nurse. I was happy to see her. "You came at the perfect time!" I exclaimed as I opened the door. "He just died."

If she was surprised at my greeting she hid it well, Hospice workers must face all sorts of unusual events. I walked her into the living room to see his peaceful body. She listened through her stethoscope for a heartbeat, for the required two minutes, but I knew there would be none. He had chosen his moment to go with perfection and grace.

The nurse was going to take care of the death certificate and she offered to call the coroner to have the body taken away, but I told her I was keeping him at home for a while.

"More than twenty-four hours and I'll have to notify the health department," she said matter-of-factly.

"Fine, do it," I replied.

Soon the house was filled again with family and friends. My step-daughter and I washed the body, wrapped it in the white linen, and with a little help moved it to a long table in a large back room that we didn't heat in the winter.

On side tables around the room we placed flowers, and candles, and incense. There were also a few chairs. Anyone who wanted to come and say goodbye to Rick's physical form was welcome. So many people came it turned into a party. People in the back room were crying and people in the kitchen were drinking wine and laughing. It was perfect.

Death is a part of life. Just as a forest could not survive and be complete without dead trees, our human ecosystem needs death too. We cannot stay static. Power, wealth, even love, must shift and move through its cycles. Like waves from the sea we each must finally reach our physical shore.

Even dead, Rick looked handsome, and he stayed that way. As one friend remarked, "he's been dead for two days and he's still the best-looking guy in the room."

At night when the house was quiet I would creep back down to say tearful goodbyes to the physical form I had spent most of my life with. I couldn't bear to part with him while his heart was still warm -- and it stayed warm for days although the room he was in was cold. Finally, by day three, although his body was still in perfect condition, it began to feel like there was a corpse in the house and I had it taken away. That is when the grief truly began.

Some people are criers, but I was never one of them. I used to think there might be something wrong with me that I didn't cry much. Was my heart not open enough? I wondered, I kind of wished I were a crier.

And then, when Rick was diagnosed with terminal cancer, I became a crier overnight. When he was in the downstairs bed because he could no longer make it upstairs, I would squeeze in next to him with my head on his chest and the tears would just roll out on their own, dampening his shirt. Before he passed, I would take walks with friends, crying the whole way, and explaining to them that this was called *anticipatory grief*, as if they didn't know. After his body was taken from the house I cried every day for months, maybe a year. I remember thinking that I didn't have to worry anymore about not being a crier. I was doing a fine job of it.

And it wasn't just the eyeball aquatics that changed for me. I suddenly felt so *delicate*. I felt like an egg that had lost its shell but was still held together by that thin membrane just under the shell. One false move and the membrane would tear and the contents would spill.

I also became hyper-connected to the supernatural. We hear bits and pieces of the stories – the things that the dead can make happen in this material realm after they have passed on; I have no idea, nor even a hypothesis, about how that happens, but I now know that it's true.

First there were the nickels. It is not unusual to find little bits of change around the house or in the street. But suddenly there were nickels appearing everywhere. No pennies, no dimes, no quarters, only nickels. I was living alone so it wasn't something someone else was doing. For instance, I used the small downstairs powder room to pee in the morning, but when I went back into the bathroom later that day there was a nickel in the middle of the floor. "Oh, you again, how interesting," I'd think. I couldn't detect any specific message connected with them. If anything, I thought they might be saying, "I am still here."

It took me years to figure out that the nickels were for his five children.

The nickels didn't really rock my world, what blew my mind was the <u>wind</u>. It started the day we buried his ashes.

I had decided to hold the ash-burial gathering in late March, a few months after his passing. It turned out to be a beautiful spring day; forsythia, camellia, daffodils and hyacinth were all blooming. One of Rick's friends, a Chinese medicine practitioner, arrived a few hours early. I asked if he would be willing to walk down to the burial site with me and dig a hole for the ashes. He carried the shovel, and together we walked the quarter mile path along the hedgerow. Dark clouds began to gather as we approach the site. I suggested a spot for the hole by pointing. The moment the shovel sliced the earth there was a loud clap of lightening. Whoa!

The wind got stronger, and it started raining, we were totally unprepared for this change in the weather. It didn't take long to dig a hole in the soft earth, and we hurried back to the house, ducking raindrops.

I thought the "party" would be ruined, but by the time the other guests arrived it was bright and sunny and warm again. It was good, as always, to eat and drink and visit; but finally, it was time to put his ashes in the ground.

One friend brought bells for everyone, and together we walked the path to the cemetery while strewing flowers and ringing bells. The sky was blue. But as we got close to the hole the strange weather repeated itself. The sky got dark and the wind started blowing. Without much delay, because of the weather, I said a few words and we put the ashes in the ground. People were hugging and crying, but then the sky started crying too. It was suddenly windy, dark, and raining. The hole-digger and I just looked at each other with wide eyes. Wow. *What?!...*

The guests practically ran toward the house to escape the storm. Then, once again, as we neared the house the weather improved. We gathered under the big silver maple tree in the yard, and friends took turns playing their guitars.

Oh sassafras, oak, pine, cherry, and maple, have your roots found those ashes yet? Is something of him now in you?

After that day, when I was in deep grief, I would often walk to visit the place where Rick's ashes were buried. I had decorated the site with prayer flags and wind chimes, and had placed a chair there too.

When I arrived I would sit in the chair and fill my thoughts with his presence. As my mind became clear and focused the wind would start to blow, flags would start flapping and the chimes would start ringing. It was as if my thoughts of the other dimension had resulted in physical manifestation. My spirit reached out, and the spirit on the other side blew back. Was it Rick or was it God? I wondered, but it felt most like Rick had become God so the answer didn't matter. The message I got was, "I am here, I am aware of your feelings."

Time after time, after time, I had this same experience. The wind blew stronger after I sat and entered that mental space. I was open to the possibility that grief might be making me batty, but I also knew that I was not alone in my experience. More than a quarter of the population believes that mental communication with the dead occurs.[4]

The appearance of the nickels only lasted for weeks, the wind phenomena lasted for many months, but it was the stopped-clocks that lasted for years. Clocks were strangely stopping for no reason. Our kitschy plastic kitchen clock that read, *Home Sweet Home,* was the first to stop. The second hand was just pulsing in place like a heartbeat. Then it was the clock in his office. Then the one in our bedroom. Yes, battery operated clocks need new batteries on occasion, but I had never had so many go, and so quickly. It brought to mind the lyrics of a song about a Grandfather clock that I learned as a child in the 1950s: *but it stopped, short, never to go again, when the old ma-an died.*[5] I wondered if there was anything to that, so I Googled it. If you do the same you will read page after page after page of stories of others having the same experience. How do the dead do it? I wondered. And why clocks?

Perhaps the message Rick and the other clock wranglers are trying to tell us is that *time is not what we think it is.* As Einstein wrote: "The distinction between past, present, and future is only a stubbornly persistent illusion."[6] British physicist Paul Davies confessed, "To be perfectly honest, neither scientists nor philosophers really know what time is or why it exists." Cambridge Professor of Astronomy, Sir Arthur Eddington coined the phrase "time's arrow" in 1927, meaning simply that physical things have a direction of change from one state to another that cannot be reversed. This is our experience of time; however, the fundamental equations of classical physics do not distinguish between past and future, they are time-reversible, time exists in its entirety all of the time.

There must be a reason why those who have passed beyond can influence clocks and *are* influencing them. But the message feels like it is written in a foreign script that I cannot fully translate. The message is something about time being an illusion, yet I am getting older every year anyway -- it sure feels like it is going in one direction. There are mysteries here, indeed, but perhaps we aren't meant to understand them.

Five years after Rick died, I was living in Colorado. You will hear more about that soon. One morning I came down to my Colorado kitchen and saw that the kitchen clock had stopped. Nothing too unusual, except it was the *anniversary of his death*. I send love, I go on with my day. They are still here with us, you know. The veil is so thin.

Being a strong Christian for many years, I believed in the tenant that went with it of 'life after death' – a belief shared by the majority of Americans.[7] But maybe *hope* was a better word for my feelings than belief. Truth be told, I was not completely convinced. As a scientist I held on to the possibility that it all ends with our flesh.

That changed for me one night when I was in my late twenties. On that evening, lying in bed with Rick in our Maryland farmhouse, while drifting off to sleep I was suddenly transported out of my body and through space to stand in front of a tree I had recently planted. After a few moments of examining the tree, I was transported back to the bed and I could feel that my spirit was separate from my body. I experienced my spirit as an arched beam of energetic light. As I prepared to re-enter my body, I realized I had a choice in whether I wanted to re-enter or not. I chose to re-enter. When I returned to my body in the bed I rolled over and told Rick,

"Now I _know_ there is life after death." I never had to hope or doubt again.

In the first year after Rick's passing, even with my firm belief in life after death, and the signs coming from him, I was still in the depths of grief. I didn't want him in that other place just beyond the veil, I wanted him here with me. I wanted to share dinners, stories of the day, warmth in bed. This life alone was empty and cold. I have heard it said that grief is love's souvenir -- it's our proof that we once loved. Grief is the receipt we wave in the air that says to the world, "look, love was once mine. I loved well, here's my proof that I paid the price." But none of that was helpful, nothing was helpful. The well-meaning sympathy cards, that tried to reassure me that I had wonderful *memories*, only made things worse. At that point I would have been happier having no memories whatsoever.

I wasn't going to end my life prematurely, but I did pray that some higher power would do it for me. As Rumi writes, "I didn't come here of my own accord and I can't leave that way. Whoever brought me here will have to take me home."[8]

Rick's children were in grief as well (four from a previous marriage and one from ours), and we were all confused about what life would be like now, and how we were going to make sense of it all. Would his death mean that we had entered a forever-time of darkness and grief? Or would we somehow, some way, all be OK again someday? There was no longer a tether to the ground. We had no idea where we would land.

One morning, while his son and I were standing in the kitchen awkwardly trying to address this new and unknown life, I looked in his sad eyes. As a mother I felt it was my responsibility to help him through, even though I had no idea how to do that for myself. "We have to bring joy back. That's the most important thing now," I said, to him -- and to myself.

He just looked at me. "But how?"

My mind cast about: *joy, joy... when, where, had I felt joy*? It seemed so far away now. Where in the world was joy big enough that it could blow through this oily dark cloud of grief? And then the words came out: "Burning Man. Let's go to Burning Man together."

It seemed like a radical suggestion for a new widow in the grips of deep grief. Amazingly, (or not?) when the mail arrived that day, along with the sympathy cards was a postcard from the Burning Man organization announcing the opening of ticket sales for 2011. This was before the days of highly competitive online ticket sales. If you wanted a ticket in those days, you could get one. It seemed like a sign from the spirit world – a place I was living in as much as I was living in this one. I signed up for two tickets for the event happening in eight months. Later, Rick's son backed out, as did another friend (neither of them had ever been before), but by then I was surviving only by following my instincts and my instincts were leading me there.

Can you see her? She is like a wild vine, struggling her way toward the light.

The Forests

You must change your life, wrote the poet Rilke.[9] I knew it already. My former life as a professor, coming home exhausted, but to a lovingly prepared home-cooked meal, nice wine, and supportive company, was over. I was no longer a householder. Instead of longing for the past, or piecing together some ill substitute, I knew I must dive into the future, I had to change my life. And a great gift from the universe had already shown me what that change would be.

In the years before Rick died, I had been traveling to the most pristine forests left in the eastern US. I was doing research for a book I was writing at the time, *Among the Ancients*. Often, I went alone, backpacking into wilderness areas and sleeping beneath remarkable trees.

Those journeys brought with them frequent mini-miracles, such as the day I drove eleven hours straight to get to a forest in Indiana, and then I pulled over to sleep on the mattress in the back of my truck because I was so tired I couldn't drive anymore. When I woke in the morning, I discovered I was *in* the forest I was trying to find!

One day, while driving my truck on a visit to old-growth forests in New England, it suddenly came to me: the last of these forests are so precious, and so rare, we need to make sure they are preserved forever. *Who is focused on saving these forests?* I wondered.

There were some effective local groups I knew of, but in other places there was no local group. What national organization was focused on this? I couldn't think of one. *And we should be protecting some of the oldest second-growth forests too*, I thought, to nudge the amount of old-growth left from one percent to... to something more. Ideally a protected forest in every county – a forest school children could visit that would never be cut down.

I pulled over to the side of the road when these thoughts started flooding through me. We could call it the Old-Growth Forest Network, I thought, almost without thinking, while gazing at the forest covered hills from the window of my truck. To accomplish this mission I would need help, I would need an organization to support me, because I would have to quit my "day job" to accomplish all that needed to be done. The whole vision came to me at once. It was ten minutes that changed my life.

When I got home Rick asked about my trip, I told him that I had recognized my next, "life's work." I was excited as I described it to him, but I noticed a dark cloud flash behind his eyes. He didn't say anything negative, but to me that cloud told of the price he paid for having an ambitious wife: the dinners for one, the nights alone. An organization like the one I was describing would mean even more work, even more travel, at a time when he was hoping I would finally be slowing down. That cloud told me the time was not right, but I never gave up the dream.

I have seen fifty-year-old redwood trees that you can circle with your thumb and forefinger. They are just waiting for the right moment, when a gap opens up above them.

Just weeks after his death I went into the Dean's office and announced my 'early retirement'. It was a wild move, for sure. The Dean, as well as my friends and family, thought I was suffering from grief and delusion. Rick's paychecks ended with his death, and now I was giving up my paychecks too, just when I needed them more than ever. Strangely, I had no fear at all. This was something that had to be done, it was something I believed in, and now Rick would not have to sacrifice for my calling.

I had gone through "the change" and the energy that formerly went to my reproductive cycles, or my husband's needs, was now available for the wider world.

Burning Man Two (age 55)

By the time I returned to Burning Man, just two short years after the first trip, I was no longer married, I was no longer a college professor, and I was no longer a virgin burner. I was experienced enough to know that I needed a camp family, a safe space for my physical needs – extra shade, a kitchen space – and my metaphysical needs. Nectar Village felt like safe space for me, how ironic!

I didn't know anyone there personally, so I reached out online and they put me in contact with the organizer of Shaman Dome – the camp in the village that gave Spirit Animal workshops. The head of the camp heard my story and welcomed me.

I no longer had any worries about the start of fall semester. In August I flew from Maryland to California and then got a ride to the Nevada desert with someone from the camp, a stranger to me, whom I had connected with by email. This time I arrived in the desert with my tiny tent and one carry-on sized suitcase.

Anything you want is at Burning Man. Many come for sex, drugs, chest-thumping electronic dance music, or maybe a combination of all of these. but I am here on a spiritual quest, a pilgrimage to bring closure to my former life.

Although I was still in grief, I had survived. I had not gone to the 'dark side.' I was still seeking joy.

I had spent the previous eight months cleaning out closets and communicating with the wind. When I got to the closet with his wedding suit, I couldn't bring myself to put it in a bag for the thrift shop. And my wedding dress? What to do with that? The answer came in a flash: I would bring them to Burning Man and they would burn in the Temple. Inside that small carry-on suitcase along with my clothes and some battery-operated lights were our wedding garments, tacks for hanging them, and the brocade-cloth decorated with dragonflies that I had purchased in Mexico.

What I didn't know until I arrived was that all of the other camp members were truly Shamans with a capital S. They were people who had devoted much of their lives to connecting with unseen forces. They came from all different places and all different traditions, but all of them were planning to live and work together at Shaman Dome in Nectar Village for a week.

There was an abundance of drums and feathers and rattles and beads. They were prepared to lead workshops and provide private healing sessions. As we were figuring out volunteer assignments – everyone was expected to participate – they looked at me and I shrugged, "I'm not a Shaman." A few wondered how I even got to be a part of their camp, but there I was.

In the end I was given the light-hearted job of being the 'line fairy.' In the midday heat I sprayed people with a cool mist of water and talked to them about what they wanted/ needed from Shaman Dome. Some newbies I steered to the huge, ornate tent filled with pillows where the *Find Your Spirit Animal* workshop was about to begin, other visitors were in crisis and needed private counseling. Some of the workshops were more advanced. One woman in line told me she already knew her spirit animal – it was a rainbow snake – and I excitedly paired her with the Shaman who also had a rainbow snake as his spirit animal. An hour later I saw them still deep in conversation, and I was pleased with my role in bringing them together.

On the first day a young woman who had just suffered some sort of personal trauma came into our camp. She was in tears and all the healers in Shaman Dome gathered around her. Some were talking, some were touching, some were using rattles, but to me the most interesting person was the man who was on the outer fringes of the group and moving his hands as if he were playing a three-dimensional air harp. I was mesmerized. As a mere 'line fairy' I didn't participate in the healing, but later I got up my courage to talk to the air-harp healer. He was attractive, with short hair and a trim figure, about ten years younger than me. His playa name was Wizard. I asked about his methods and he explained that he could see energy. "I don't know if I'm doing any good," he said humbly, "I just see the energy and move it around." He told me he lived in Mexico City, and I mentioned that I visit Mexico every year. We exchanged contact information and he said he'd come visit me the next time I was there.

Anyone who has heard of Burning Man has heard of the made-of-wood man that gets burned (by the way, this can be a symbol of anything you want it to be). Forget about The Man, I wasn't really interested in that anymore. I was here for the Temple. While both of these structures are huge, and they are both unique every year, and they both get burned (The Man on Saturday night and the Temple on Sunday night), beyond that they are very, very different from one another.

Artist David Best constructed the first Temple, in 2000, to commemorate the death of a friend who had passed away in a motorcycle accident on his way to Burning Man in 1999. Every year since then the Temple has been different, designed and built by volunteers. This year it was named the Temple of Transition, and it was the largest Temple ever built.

Right after I arrived I went to visit the Temple and it was more wonderful than I could have hoped for. It had five minaret type structures, representing different life phases, surrounding a central tiered ornate tower one-hundred-twenty feet tall. Inside the tower the ground was shady and cool. Built into the structure was a type of mechanical music box that played Balinese Gamelan music. Lovely, otherworldly. Some of the visitors were meditating, some were dancing. One woman in the center of a circle was moving her lithe body to the sound of the bells so slowly, so fluidly, I couldn't help staring. Another young woman, dressed like she had just stepped out of a Sulamith Wülfing painting -- complete with a hooded garment made of white fur -- stood and started singing a wordless song that sounded like it was composed by a choir of angels. I wasn't in Kansas anymore.

The Temple is a mystical community space where important passages are recognized. Things are let go of and are honored in this space where silence and tears are more common than laughter. People hang things they have created honoring their journeys. Names and messages to the universe are written in colored ink on the boards that will soon burn. Cardboard posters with photos taped and glued to them are everywhere. Already, on this first day, there were remembrances and small containers (likely containing cremains) nailed, pinned, taped, and tied to the structure. This poster, here, was made for someone's dear friend who *passed over the rainbow bridge* this year. The photos taped to it are of a lovely young woman we'll call Rebecca. Rebecca with her bright smile, Rebecca with her best friend (the poster maker), Rebecca with her dog. The words on the poster reflect what this bright spirit meant to her friend and how hard it was to lose her. I'm starting to cry and I don't even know her. Next to that is a sign someone has hung celebrating their journey to sobriety. Alcohol had taken so much from their life, they had written, but this was the year they finally found the strength to break that bond and release their dysfunctional drinking. Prayers of gratitude were offered along with apologies to all they had hurt along the way. *Right on brother.*

The Temple is like a forest of human emotions. Layer after layer of things to explore and feel. Always changing. Every community needs a space like this perhaps as much as they need a forest. A place to share, to honor, to let go.

A photo album around the corner was placed by a woman who got divorced this year. There were pages celebrating the love that they found, and all the many special times they had. There were wedding photos, photos on mountain tops, photos at parties, and photos with babies. With the imminent burning of all this she honored not only what they had shared but released whatever hold it had on her. She wanted to move into the future free of regrets and second guesses.

People had hung all sorts of things: stuffed animals, pieces of jewelry, a military uniform. There were hundreds and hundreds of these memorials. It was impossible to see them all. But around the corner one caught my eye. It was handwritten on the structure:

Loved fully
Grieved deeply
Let go completely

This brought me to a sudden halt. <u>Yes</u>, this was the answer, this is what I wanted for myself. Let this be my mantra.

My pilgrimage with the wedding garments would occur at dawn the next day.

The following dawn I dressed in gauzy white and carried our wedding clothes folded inside the dragonfly cloth. The Mexican Wizard and a Native American Shaman from our camp accompanied me. No bikes, just footsteps in the white dust, while we made our long, silent, journey across the vast playa under a pink sky. When we entered the portal of the Temple they left me alone.

Already at this hour there are people sitting quietly and meditating or sleeping, and others standing and reading the offerings on the walls. I look around for where my offering should go. I look up, up, and there on the second floor of the central portal – the heart of the temple – is a space calling to me. I climb up as if in a trance. When I get to where I can access the space I sit and slowly open my bundle. The tears began to flow. Right there is the simple white muslin dress I wore, with baby's breath flowers in my hair, over thirty years ago; there, right there, is the white cotton suit he had worn while holding my hand, nervously patting it, as we approached the wedding guests waiting under the shade of the giant cedar tree.

Who were we then? I tried to remember us. In the decades since that formal union we had certainly changed, but the soul connection had never faltered; even now, I felt connected to him. But I also knew that while I would always have a spirit connection to Rick, I must let go of our physical connection.

There was a ceremony when we joined *(the two shall become one)* and now there would be a ceremony for the unjoining *(the two shall become one)*. A different one. I must let go of the marriage if I were ever to be a full person again. Saying good-bye to him had been hard enough, and it was still ongoing, but now I must also say good-bye to that wonderful time in my life when I was part of a couple. A couple who created and raised a child together, a couple who threw dinner parties together, a couple who helped each other carry groceries from the car to the house, and who shared the same bed (...*loved fully*). These were the memories and the prayers flooding my being as I unfold the garments and, using the tacks and a small hammer, I begin tacking the dress and the suit jacket to the wooden beam where they will hang down into the central space. It is a heart-crushing good-bye. I feel as if I am being torn in two, and energetically I am. After I tap in the last tack I break down into a deep mournful cry.

It is a powerful, good, grieving (...*grieved deeply*) the kind that causes your shoulders to shake and your nose to pour mucus. When I was finished I wiped all the liquid from my face and turned toward the sun, now above the horizon. The warm sun and the cool desert breeze brought a feeling of peaceful contentment. And then it happened....

In this desert where I had never seen a single living thing besides humans: not a single weed, not a single bird, not a single mosquito or even a gnat; there appeared right next to me, a *dragonfly*. At first I assumed it was dead, brought here as an offering, and just blown in my direction. That would be special enough, since my past shamanistic journeys had already shown me that I had a special connection with dragonfly -- it had been my spirit animal for years. But look! It is *alive*. As if by instinct, I reach out my hand, and the fantastical insect climbs onto my finger. Now I am crying again, but my heart is singing at the same time! I hold the spirit creature up and stare deeply into its pale brown eyes, it seems to be staring back into my light-blue ones. When it finally flies away I am left with no doubt that it was a messenger from a different realm.

My feet were light as I made my way back to camp. *A dragonfly visited me in the central tower of the Temple at Burning Man at the very moment I completed the mission of my pilgrimage.* I was blessed.

I returned to the Temple every day. I asked numerous people if they had ever seen a dragonfly at Burning Man and their answer was always "no." Every time I approached the Temple I felt the tiniest inward gasp when I saw the garments again. *His* body was once inside that jacket -- *her* body was once inside that dress. *Who were they then?* I tried to put myself back there. I was nervous, too, on our wedding day, but looking back I had definitely made a good choice. We were two who had surely become one.

Our dear redwood-wedding friends from 2009 were back at Burning Man too, but they were far away on the other side of the playa. Fortunately, the Shaman Dome camp brought bicycles we could borrow. I chose a little one no one else wanted, it had a banana seat and ape handles – more of a kid's bike really - but I was grateful to have it to cover the vast distances. One day I pedaled across the playa to spend an evening with my young friends, and when it got very late we slept together on a mattress in the back of their school bus.

When the sky turned gray, announcing the imminent sunrise, I slipped out of our shared bed and gathered my pile of belongings. I was still wearing the little silk slip that I wore as a dress on the playa – so no worries there. Outside the bus I found a chair where I could sit to dust off my feet, put on my socks, and pull on my black cowgirl boots. (You can go around buck naked on the playa if you want, but keeping your feet away from the alkaline dust is important.) I wound a colorful scarf across my shoulders, replaced my baubles and beads, put on my hat (circled with battery-operated lights), swung my leg over the banana seat on my little bike, and pedaled off. It must have been close to five o'clock.

Burning Man is active twenty-four hours a day. Some people like the daytime things there best, and some like the nighttime things best, but five a.m. is the quietest, emptiest, time there is and for that reason it is my favorite time. I was riding along foggy-headed and thirsty when I passed a bar camp with a sign advertising *Bloody Marys*. Hmmm. Yes, I could do that. I swung in and there was a woman behind the bar happy to make one for me.

Keep in mind that there was no delivery truck that was bringing in tomato juice, no distributor bringing vodka, every last thing, including the bar itself, had to be hauled out to the desert and would have to be hauled back home – covered in dust.

It was the best Bloody I've ever had. (A drink with ice in the desert is a magical thing.) The only other person on my side of the makeshift plywood bar was a shirt-cocking forty-something-year-old guy. Shirt-cocking is a fashion choice unique to Burning Man, as far as I know. Most people there wear as few clothes as possible in the daytime, but the intense sun beats down and you need protection to prevent burning, and maybe you feel just a wee bit too modest to have your little man just hanging out there – so a longish shirt with nothing under it is a functional wardrobe choice.

I had never seen these two people before, and likely never would again. The guy and I started chatting. He was open and friendly, and the conversation turned to body ornamentation. I had a rare opportunity to ask something I'd occasionally wondered about: penis piercing. Why that small metal ring I'd seen near the head of the penis on some men? Seems to me it would interfere, but what do I know? So, over my super delicious, free, five a.m. Bloody Mary, I'm having a conversation about intimate body-modification with a half-naked stranger.

I'm not sure exactly what I asked him, but he got very animated. "Oh cock-rings!" he exclaimed. "I wear one every day. I can't believe I don't have one on today of all days when you asked about it."

Well, I think, at least I asked the right person. "So why do you wear one?"

"You know," he says, gesturing to his lower half, "it's for *presentation*. Like, the same reason you'd wear a push-up bra. Here, let me show you," he says as his eyes swivel around the camp bar. Thinking to himself...hmmm...no, no, no.

Then he says to the woman behind the bar who made our drinks– "hand me that plastic ring." He was referring to the small ring that's left behind on a plastic gallon jug after you unscrew and remove the cap. "Yup," he says, "this will work, let me show you. First you feed the balls through one at a time....." He was as casual and unselfconscious as if he were teaching archery.

Believe me when I say that this was *not* sexy, but it was very, very, interesting. I like unusual things: I'll taste the fried grasshoppers in Mexico, or the eat the jellyfish in Thailand, or sign up for the zipline excursion where I'm vacationing, but this was about as unusual as it gets – watching a stranger in a strange land thread his balls, one at a time, through a little plastic circle for no reason other than educating me.

After both testicles made it through, his penis was threaded through the space between them, so all three appendages were squeezed together at the base and projected a little more from his body than they normally would. "See," he said, gesturing at the completed assemblage.

It was then that I realized we were talking about two completely different things. My original wonderment was about *pierced* penises, not about cock rings at all. In fact, before this demonstration, I didn't even know there was such a thing as a cock ring. But he was so earnest, and so pleased that he could be helpful, that I couldn't bear to tell him there had been a miscommunication.

When my drink was finished, and I was saying my goodbyes and getting ready to get back on my child-sized bike, he reversed the process, first unthreading the penis, then each testicle, and then he handed me the plastic ring.

"Here," he said. It was a gift. One of the strangest gifts I've ever been given. I wasn't certain how I should react, so I just slipped it onto my thumb as I climbed on my bike.

"Thanks!" I called back, as I waved my arm high and pedaled away.

I actually kept that plastic ring for a while. That's Burning Man, can't make this shit up.

One day I am having mystical experiences I'd call holy, and the next day I'm living as raw as it gets on the physical plane. I look at myself here and I see my nature: forever curious, forever bold. These are good traits for a scientist, no wonder I went in that direction.

There were more adventures along the way, but I made it back to Nectar Village in time for rehearsal.

Someone at the acro-yoga camp had offered to teach us the Balinese monkey chant; the rehearsals were open to anyone interested and they took place every morning. By late in the week over a hundred people were packed under the open-air yoga tent chattering like monkeys in a call and response.

Rick and I had been to Bali in the early nineties. We had gone to the temple dances and watched the shadow puppet plays, we watched the graceful young girl dancers with their fingers bent back into extreme positions and their eyes shifting this way and that, we heard the men, wearing nothing more than black and white checkered waist wraps and a headband, playing their percussion instruments and chanting. These events were meant, most literally, to *enchant*. The performers were working to bring God down to earth and literally into their bodies, sometimes in the form of an animal.

But this God was not a single remote being…this God they were entraining was more like the sun, it could pervade anything, and it brought contentment, happiness, and beauty to those whom it visited. The Balinese believed that this god-like power could also be harnessed by those who were deceased, and it gave them the ability to be a positive force in the lives of those they had left behind. For that reason, in Bali, deceased family members are honored and asked for help. Every yard has a small shrine to the ancestors where rice, fruit, and flowers are left, and favors may be asked.

This way of life is pre-Hindu, pre-Buddhist, pre-Islam.

On a few occasions, while visiting the temples, we got to witness the Ramayana Monkey Chant, also known as the Kecak. In this performance a large group of men, sometimes as many as a hundred, *become* monkeys getting ready for battle, and they repeat the same syllable over and over: "chak, chak, chak, chak." Sometimes soft and sometimes loud, sometimes slow and sometimes fast, all the while making arm movements to match the rhythm. Different parts of the group take over at different times, creating an effect like a hundred people singing *row, row, row your boat* in a round with five different parts, only way more interesting. I was captivated by watching the performances in Bali, but I never dreamed I would be a part of one.

But here I am! I am in a group of a hundred people in Nectar Village chaking away. We could have been anywhere on this vast playa, drinking, dancing, fucking, sleeping, but here we are, together, bringing God down to earth.

One day, at rehearsal, it was announced that we would be performing the chant in the Temple, at midnight (!). As I made my way to the Temple, once again, and took my seat on the floor of the central dome, with my dress and his jacket hanging in a place of glory above, and the walls covered with the deepest love-filled thoughts and prayers, with gamelan chimes ringing, and people chanting, I knew that this was exactly where I should be. I had followed my instincts, they led me here, they would lead me onward…chak, chak, chak chak.

The Buddha said, "If you cannot find friend or master to go with you, travel on alone –like a king who has given away his kingdom, like an elephant in the forest." I was slowly getting used to these solo steps.

On Sunday, the last night of Burning Man and the night of the Temple burn, I arrived clean, anointed with oils, and dressed for a ritual. Around my shoulders I wear the dragonfly cloth, in my hand is a bundle of sacred white sage. I come early so I can sit in the front circle, in a place where I can see through the entranceway to our wedding clothes. Soon thousands more people arrive. They all sit in silence out of respect. It is very, very different from the burning of The Man the night before. I light the white sage and the smoke wafts out over the gathering crowd.

After a while I can see the fire crew inside the Temple. They start a very small fire in the central sanctuary. Suddenly there is a whoosh of blue flame. The muslin gauze of my simple wedding dress is the first thing in the memorial structure to invite the fire. And in a flash it is gone, and then his jacket, and then all the paper notes tacked to the walls, and then the wooden railings, and then the floorboards, and then even the roof; and finally the entire structure is consumed by huge dancing flames.

The heat is so intense I have to use the dragonfly cloth to shield my face. And in the light from the flames the thousands of people who had honored someone there were reminded that it is good to care, good to feel deeply; but, finally, this physical plane is just a mirage and, like the memorial structure, it will soon be gone.

My former life, too, was being swept away. Even the forests I care about so much are sometimes consumed by the flames and start afresh. This is the way of life – always starting over.

When the flames died down to a smolder the observers got to their feet and walked clockwise around the burning debris. I joined in the circumambulation and walked many times around. As I walked, I reflected on my pilgrimage. First the months of anticipation, the preparation, the long flight, the drive to the desert, the walk at dawn carrying the wedding garments, the dragonfly; and now this circle walk with a community of other walkers. I had completed what I had set out to do, and it turned out better than I ever could have planned or dreamed. Each step of the journey, both literally and figuratively, felt like a pilgrimage toward wholeness and happiness. I had a long way to go, but I was heading in the right direction.

Well, all that was super powerful and intense, but then life, trickster that it is, had a few surprises for me: the Yang to the Yin, the Cow to the Cat.

When it was time to think about leaving Burning Man I realized that I didn't have a ride out. The warm and gentle woman I had met through the ride share board, who had given me a lift from California, had left a few days earlier. I made many inquiries, and at last I found someone with room in their vehicle; the camp cook agreed to give me a ride back to San Francisco on Monday. On Tuesday I would catch my flight back to Maryland, back to my purring cat, back to the flocks of migrating birds, back to the place where Rick's ashes were buried and the prayer flags waved in the breeze.

Burning Man 2011 was officially over. The dusty migration back to the 'default world' had begun. Everyone was packing up and heading out. I crammed my dirty, powdery, clothes into my carry-on sized rolling suitcase. Then I folded up my small tent just before the camp's shade structure came down. All around me there were good-byes as the Shamans rolled away. Finally, I was left alone on the dirt with just a small suitcase and the bundle that contained my tent. For a moment I was washed with pride: *Look at me, I have done it. I came to Burning Man all by myself. Look how lightly I travel. I am amazing aren't I?* But soon enough I was over that conceited back-patting and I was just waiting to leave. The playa bars were all gone, there was no shade, no food, and nowhere to even sit except on my suitcase. As the sun lowered toward the horizon my mood lowered too. Then I got word that the cook was not leaving that day after all. He had decided to stay a couple of nights with his new lover now that all his other responsibilities were over.

What was I going to do now? Any trace of smugness had been instantly washed away. What was I thinking, leaving myself this vulnerable? I needed to start being a bit more careful now that I was alone in the world.

Everyone at Shaman Dome camp had been kind and supportive, but there was one person I avoided. She was almost a caricature. She had short, dark hair, was wide across the middle, and wore lots of leather and chains. She taught a workshop on *Spanking*, and she walked around with her special tool which resembled a leather flyswatter. I stayed far away from her. I attended many of the other workshops in our camp, but not hers. I couldn't imagine how she became a part of Shaman Dome, but the unusual eros of Burning Man was alive in our camp too.

The sun was starting to set and only one piece of our camp was left. It was one of those standard U-Haul trucks with a front cab and a big box shape in the back. Many burners rent these trucks to carry all their gear to the playa and then they also sleep in the back. Guess who was staying in that truck? Guess who was kindly offering me shelter? The *spanker*!!

Oh, the irony. The one person I was afraid of, the one person I had avoided all week, was the one person who was now offering to help me. I accepted her offer and shared the back of the truck with her that night. Turns out she was kind and respectful, and I was grateful. As my friend Kath used to say, "another fucking opportunity for growth."

In the end I hitch-hiked out, a hot and sobering experience. I watched the formerly generous nature of the burners evaporate as they headed toward the exit gate. Vehicle after vehicle passed me by with my cardboard sign announcing my destination –" San Francisco." I was starting to be concerned about heat stroke when I was finally offered a ride. I think of myself as brave, but over and over again I get tested: *How brave are you really?*

The dragonflies have continued to visit me. They are definitely my 'spirit creature' and everyone in my life knows it. I have been given many gifts of items with dragonflies on them. You should see my jewelry!

Live dragonflies visit me constantly, but especially when something very positive is happening: walking a path down to a secluded beach in Hawaii, watching my grandson learn to swim, sitting on a beautiful rock outside my new Colorado home, in the yoga studio in Mexico.

Always they appear when I am feeling especially blessed. I would have thousands upon thousands of examples if I ever kept count.

What do I make of their presence in my life? My answer comes from my gut and not my head. Even though I am a PhD scientist I believe there is something of another dimension to their appearances. I get the feeling that something of Rick has sent them, or is in them. The message they bring, is clear: "*See,*" they say, "*see,* everything is going to be alright, better than alright. We are watching. We have helped make this happen. We have smoothed the way for you." It is always a wonderful blessing to receive, and I am grateful.

I am not the only one who feels messages from the other side in the beings of this world. I know one woman who experiences her dead father in the appearance of red birds, like cardinals; another friend gets sent butterflies by her grandmother; a man who was also a recent widow pointed out the hawk that came to watch over his daily walk --every morning since his wife died. It's fascinating to me that these "messages" are usually carried by winged beings. Angels have wings too.

Zipolite Two

Slowly, over that first awful year after losing Rick, I came to recognize that I was not the only one feeling the loss. Our friends in Mexico deserved to hear, firsthand, the story of what had happened. So, almost exactly a year after he passed, I returned. I thought this would be the last time I'd visit. I would simply tell our friends the story of what happened, together we'd remember Rick, and I would close that chapter of my life. I saw it as similar to the honoring/ closing work I had done at his home-wake, at his memorial service, at his ash-burying, at Burning Man.

But something unexpected happened once I got there; my heart felt uplifted and at the same time the land energies grounded me. Higher, happier, more solid, all of it at once. I had roots here. There was no reason to end one of the best things in my life. I would continue going to Zipolite, alone if necessary.

The following year I arrived in Zipolite on December 24th, the busiest time of year (with Easter a close second). I had no room reservations because I planned to stay with my friend Peter Pan. But when I arrived Peter was moving out of his place. The property owner had just kicked him out (maybe, partly, because he let friends like me stay?). My situation wasn't something he needed to worry about – he had his own worries that day.

I left my suitcase in his yard and went walking to look for a room. In those days I had no cell phone service there, and even if I did have it my Spanish was pathetic over the phone. It was my feet or nothing. Feeling a bit like Mary, I visited hotel after hotel asking about a room for the night in my broken Spanish. They were all full. Finally, I went to the one place I said I'd never stay. The place with filthy shared bathrooms where many of the toilets were missing seats, the place where the owner was always flirtatious, in a creepy way. But I was desperate, so I went to talk to him. Even he didn't have a room for that night, but he had one for the following night.

Now, being stuck without a room in Mexico is not the same as being stuck without a room in New York City, or even Maryland. The climate in Zipolite is gentle, and generations of Mexicans (as well as international backpackers) have survived just fine sleeping outdoors in hammocks. I went back to

Peter's place and asked if I could sleep on a hammock in his garden.

Connie, a Mexican native, had been living at Peter's place, so she had to leave too, but she would not accept me sleeping out in a hammock. "You are coming with me," she said. And she led me to her new casita, a small place with just one bed -- a double mattress on the floor. "You can stay on that side" she said, as she pointed, and I was filled with gratitude for this generous and loving person who literally took me into her bed when I had no place else to go. It was a real-life Christmas story.

The next day I moved into my rustic room in the dumpy motel, and the Wizard from Burning Man was due to arrive the day after that.

On the playa no one talks about their 'default' lives, so I had no idea what the Wizard's life was like, or if he even had a job. He could have been homeless. All I knew was that he could see energy and I felt a strange attraction toward him. When he showed up at the beachside bar, on the appointed day and time, he looked a bit out of place with his polished leather shoes and his rolling luggage. No homeless backpack-toting Shaman was he – turns out he was a successful businessman with a chain of stores and the latest iPhone in his pocket.

We found a table for two and got to know each other over cheap bottles of Mexican beer. The conversation went deep very quickly, and he asked to see my palms. I have the typical quarter-circle line around my thumb pad and two partial horizontal lines above that. When he showed me his I had to stifle a gasp, the lines on his palm looked like an asterisk, or a starburst, all radiating out from the center. His fingers were different too – they were unusually small. We sometimes joke about people being aliens, but I could easily be convinced that he was one.

I was embarrassed to bring him back to my rough room, but I just tried to relax into the inevitable conclusion of the evening. It was not something we discussed, just a direction we were both going. For me it was a direction of healing. I didn't have any notions that we would ever be a 'couple.' But I did have an idea that I should move forward, I should be 'in my body' again with a man – and a man who could see energy seemed to be a good choice. He wanted me, too, and it's a good thing the little room we were in had a second bed because the first one literally broke under us and collapsed on the floor. Laughing, we just moved over to bed number two.

Our time together was unusual. Beyond the first evening there was almost no 'getting to know you' chat. When we walked up the beach, maybe to get breakfast, he would walk a few paces behind me. Doing energy work? I don't know. When we'd sit to eat there would be a comfortable silence. Waiting for the food to come we'd meditate, looking out at the sea. At least four times a day we found ourselves pausing for deep meditation. Sometimes he'd give me a few suggestions like: "be aware of every single wave all the way to the horizon" or "hear every sound there is in this moment." Once he told me to stand with my legs wide facing east and to welcome the energy into my body. These suggestions and directions weren't given in a demanding, or 'better than you' way, and I was grateful for his gentle guidance. I was getting free one-on-one energy healing in my favorite place on the planet – priceless!

If you have studied eastern religions at all you have heard of kundalini energy. It is something difficult to define or control, and volumes have been written about it. I had been hearing about it for years but had never chased after that mysterious experience. Yet one dawn when I woke with the sleeping Shaman next to me, I felt a rush of energy, like a breeze, moving up my spine and I knew right away what it was.

After our night in the cheap motel, another Mexican-American friend found room in one of her rentals. On New Year's Eve, saying good-bye to 2012 and welcoming 2013, a group of us sat at candlelit tables on the beach and savored a dinner of freshly caught fish grilled in front of us on an open fire. The wine and beer were generously shared. After eating, the Wizard and I excused ourselves to sit on the beach near the fire. As usual, we were silent, staring into the flames. Elsewhere people were partying to loud music or gathering in cities for the ball drops at midnight. But we were deeply engrossed in the moment with nothing but the sound of waves and the sight of flickering flames.

Like many shamans around the world, the Wizard occasionally used a little 'plant medicine.' He described it as a speed bump, just to lift him easily into another plane of consciousness. I wasn't sure exactly what was in that pipe he passed to me over the fire, but I took a toke anyway.

We wandered down to the water's edge and sat meditating on the moonless night-sky filled with thousands of stars. Against that dark backdrop a vision emerged: a male and a female figure, like gigantic golden Balinese shadow puppets. They moved in tandem like a Temple dance staring Shiva and Shakti. I was enjoying the vision and was not at all frightened. "What are you seeing?" the Wizard asked. For some reason I chose to keep my golden figures a secret.

I didn't feel *used*. If anything, I was using him. And I returned home to Maryland a different person than the one who left. It wasn't just the ocean, or the meditation, or the sex – though all of that was nourishing. I am convinced it was the energy work. Maybe coupled with the close of the long Mayan calendar? When I asked Sri Juan about 2012 I could have never guessed what it would bring for me. I had just ridden the biggest wave of my life.

By the time I returned home to the farmhouse Rick and I had shared for so long I was grief free – a miracle I never expected.

Love fully
Grieved deeply
Let go completely

My prayer had been answered. My tears for Rick had dried and only the deep love between us remained.

People are So Generous

When I decided to quit my safe university teaching job and start a nonprofit, everyone thought I was taking a big risk; but, as Joseph Campbell says, when embarking on such a journey, "you throw off yesterday as the snake sheds its skin." And that is just what I did.

I didn't spend much time at home those blue-green days. My daughter had her own single-successful life happening in another city, and there was no one waiting at home for me. I could be gone, gone, gone, without any guilt. I was still feeling delicate, but I started giving talks about saving our forests, building the Old-Growth Forest Network, and trying to raise funds -- even before we were awarded our charity status.

A friend agreed that her organization could hold any money donated to OGFN until we were launched. "People are so generous," she said, "we get checks every week even when we don't ask." I hoped she was right because this organization was not going to survive otherwise. We started with a balance of zero.

I had been on a nonprofit board before, but that was the extent of my experience. No longer was I collecting data for scientific papers or grading student exams. Now I was looking for board members, opening a bank account, renting a post office box, and making a website.

I said yes to whatever invitations came in. In between speaking out for threatened forests and fundraising, I was also growing our Network of forests and showing up to dedicate forests all over the country. Sometimes I spent weeks on the road – occasionally sleeping in my car if I only had a few hours to rest and it seemed a waste of money to get a hotel room.

It was hard work, and I never could have survived it, except everywhere I went the most positive, energetic, people showed me the most beautiful forests. And not just the lovely eastern forests, like those in Pennsylvania and Ohio, I also got to hike through the western ponderosa pines, the redwoods, and even the sequoia forests. Sometimes the very same people showing me the forests were the ones who had helped to save them.

Eventually I came to realize that connecting all these marvelous forest-loving people together was as important as creating the Network of accessible, never-logged forests. Amidst the rocks, the moss, the fallen leaves, and the tree trunks, we celebrated the forests together. We were building our own ecosystem. Some of the characters were like trees, some like fungi, some like squirrels, and some like bumblebees. All with their own role to play. All motivated by the beauty and wonder of our living planet.

My mother was worried about her now solo daughter out on the road, "God knows where." But I felt that He/She/It *did* know. I felt protected by a magical net. I never had so much as a flat tire or a fender bender. As a child I had a painting in my room that depicted a guardian angel watching over a young child. When I was young it seemed like a delightful fantasy, but now it felt real. I felt watched over, protected, and I headed out on thousand-mile road trips with zero fears. And every time I would return home healthy and whole.

I frequently camped or stayed in private homes to save money. In Virginia I was invited to stay with people I had never met (there was nothing unusual about that anymore) and when I walked in the door there was a ceramic vase with Rick's face on it! My hosts had never met Rick, they had no idea whose face it was, they just saw it at a craft fair and liked it.

Years ago, a potter-friend had been experimenting with face casts, and we both volunteered. I have a vessel with both Rick's and my face on it, but apparently the potter had made other things with the casts as well. That is the factual explanation, but you and I can both see through that to the mystical explanation.

One of the audience members at my Virginia talk donated a thousand dollars. This was the largest donation we had ever received, and it felt like a turning point. Little by little we got more thousand-dollar donors.

The idea to create an Old-Growth Forest Network was blossoming like a magnolia flower, but the work was unsustainable on a personal level. I desperately needed another person to share the workload. Although our funding base was growing, I have always been very fiscally conservative, and it didn't seem like we had quite enough to comfortably bring on another employee.

Then one day the phone rang, and the youngish-sounding woman on the other end was saying that they had a family foundation, and they would like to support my work. What did I need? I knew the answer right away, as I had already discussed it with my board.

"We need another full-time employee."

"OK," she said, "I think we can do that."

We discussed a few details then I hung up and did a happy dance. *People are so generous!* I couldn't wait to tell the board.

But when I told the board they said that that was just enough for one year, we couldn't expect to hire someone good with only one year guaranteed. I should go back to the generous funders, they said, and ask for a commitment of that amount *every year for three years.*

Gulp.

I did it. And they said yes.

It felt as though the universe was providing. People are so generous!

Now we have four full time employees and four part-time employees. I feel deep in my soul the difference our work has made in the world. Forests are still standing because of the Old-Growth Forest Network. Even the current president of the United States, Joe Biden, is talking about old-growth forests and how we can protect them.

I was gifted with a wild mission, and the universe and I rose in tandem to see it become a reality.

Love/Hate/Love....?

My friend Linda says sometimes the truth is long, and complicated, and contradictory, and it is very 'unladylike' to tell it. Which is precisely why it needs to be told.

This is not a feel-good story, it is a hard story. More than one person has suggested I remove it from the book – including the main character, Jerry. This chapter was originally titled *Love/Hate*, and then I changed it to *Love/Hate/Love,* briefly, before I considered adding another *Hate* on the end.

I was introduced to Jerry when I was in New England looking for forests for the Old-Growth Forest Network. The person who introduced us said Jerry was a philanthropist and might be interested in supporting the Network. I was interested in meeting anyone who might be useful for my forest-saving work.

I was not attracted to him at first sight. After he arrived at my host's house and poured himself a drink, he turned to me and asked, "So what's your story?"

"Oh, I'm just a recent widow trying to bring joy back to my life," I replied, knowing this was an unusual response, and nakedly honest. It was the type of comment that would shut most people down, but Jerry didn't skip a beat.

"Well, how are you doing that?"

Without hesitation I responded, "Like right now, in this moment."

"Oh, you're good, he said with emphasis on both the *you're* and the *good*.

And in that quick encounter I felt seen. There was a little spark. We didn't have much time to visit on that trip, but when he invited me to his Colorado home I decided to go.

I showed up in Colorado five months later for a ten-day-long first date. I had changed so much in those five months! My heart had been blown wide open by the Wizard. And following that special time in Mexico, I travelled to Maui to give a talk alongside poet WS Merwin. It was a time filled with magic. Everyone I met seemed to give off electric energy, and also respond to mine. Love was, literally, in the air and I could still influence the wind. That was the woman Jerry picked up at the airport.

When we got to his house and had our get-to-know-you chat the first question he asked was if I meditated.

"Oh, yes" I replied.

And it was decided that we'd meditate together every morning.

He had a lovely home perched high on a ridge. In the morning I'd meditate facing the rising sun. I could feel some sort of energy, like an electrical charge, coursing through my body. Afterwards we'd have what Jerry called, "dharma talk" -- a sharing of our meditation experience. Few conversations are more intimate than the sharing of that other reality, a reality void of thoughts but filled with possibility.

By the time I flew home we had been intimate in many ways, but neither of us discussed the future or professed to be interested in building a committed partnership. Yet the phone calls and the long-distance visits continued, and thus began a the most frustrating relationship of my life.

It was frustrating for him, too, I'm sure. He admitted that he was wanting a cute little hippie-chick. And, as I came to realize, it would be best if that cute chick would roll her eyes at his distractions and deceptions and say, *whatever*.... I was not that person.

He was almost a decade older than me, but I was used to having an older partner. The most marvelous thing was how perfectly our *appetites* matched. And not just our appetite for food (local, organic, please, no factory-farmed meat), but our appetites for everything else. Alcohol? Yes, please but not too much. Socialization? Yes, please, but not too much. Silence, Yes, please but not all the time. Dancing? Yes, please. Hiking? Yes, please. Saving land for plants and animals? Yes, please. Physical intimacy? Yes, please. Neither of us owned a TV. When we were together the days and nights rolled comfortably by – until they didn't.

While Jerry didn't exactly invite me to move into his house in Colorado a few years later, he didn't resist when I told him I was going to. I moved out of my Maryland farmhouse, shipped my car and favorite belongings cross country, and flew with my cat to Colorado. Everything else I gave away, donated, or sold.

I still ask myself why I did that when I already knew there were problems with our relationship. My only conclusion is that I felt it was the next step in my healing. It was time for a change, any change, no matter the cost. I didn't want to be one of those widows who clung tightly to the past – yet there I was, years later, still in the Maryland farmhouse, still sleeping on 'my' side of the bed and hanging clothes in 'my' side of the closet. It was time to shake all that off I reasoned -- even if my methods were not sound.

I fixed up my new mountain-top room with original paintings and oriental rugs, a large antique wooden desk and a brass headboard. My clothes went in my first ever walk-in-closet and my favorite glasses and dishes got put in the kitchen cabinets with Jerry's things. I was a Colorado girl! Would I be happy here? Sadly, the answer was no.

After I set up my bedroom and had all my things unpacked and put away, I turned my attention to the garden. I love gardening, and I was looking forward to making the yard beautiful and abundant, but then Jerry told me I couldn't move certain plants because his friend Sadie had put them in. That dampened my enthusiasm -- and brought up anger. Was this going to be my home or not?

The red flags were everywhere, but I had no practice in seeing them. I had been with Rick since I was nineteen and never knew a thing about all that. Today, as a result of my relationship with Jerry, I know about red flags, attachment theory, childhood trauma, narcissism, people-pleasing, and boundaries. Now I read books on relationships and listen to podcasts about relationships. I keep learning, keep trying to figure out where we went so wrong.

One of the first red flags appeared early in the relationship, before I moved in, when I traveled with him to his hometown and we stayed in a cozy little bed and breakfast. It had been a long day, but we made sweet love under the covers. As I drifted off to sleep I murmured, "Goodnight, I love you."

We had never used the L word before, and, honestly, I didn't even think about what I was saying. It was a habit from all those years with Rick, and it just slipped out.

"What did you say?" asked Jerry.

"I said, 'Goodnight, I love you'" I repeated. I was willing to fess up that it had slipped out if he questioned me about it. And if he told me he didn't feel the same way about me, well that was fine too. Honestly, I was trying, but I wasn't sure I was in love.

But instead, he said, "Love is just a word used to control people."

What! "You don't believe in love?" I was more awake now.

"In EST training in the 1970s Werner Erhard told us the concept was just nonsense, an illusion, only used to control people."

"Well, I feel very, very differently. I feel that love is real and it is one of the strongest forces in the world."

I didn't have holocaust survivor Viktor Frankl's book, *Man's Search for Meaning,* nearby, but if I did I might have quoted from it:

> *...I saw the truth...The Truth – that love is the ultimate and the highest goal to which man can aspire.*[10]

Ah, but what good are all these words to someone who doesn't feel it? As the poet Rumi writes:

*If you want to expound on love,
take your intellect out and let it lie down
in the mud. It's no help.*[11]

Why was I trying to create a relationship with someone who didn't believe in love? Why didn't I say good-bye at the end of that weekend? Perhaps, I was convinced that with *me* it would be different. Maybe I kept going just to prove to him that he was wrong about love, and I was right.

Our problematic situation, you see, was coming from both sides. He was telling me who he was, but I was slow in believing it.

Eventually I did fall in love with him. I was *there* with him, in that special emotional place, and ready to stay with him always. Hormones had a lot to do with it, but that's why we have those hormones. On this one particular night when Jerry was inside me I wanted him to stay there forever. I felt us cleave together. For me it was our spiritual honeymoon and I wanted to continue the closeness the next morning.

But in the morning Jerry was getting dressed saying he had to run to the pharmacy. I pulled on a robe. As he sat at the bottom of the stairs putting on his shoes, I slid my arm across his shoulder and whispered in his ear, "You're coming right back?" I was shyly trying to communicate what I was feeling: *I want to get back in bed with you.*

"Well, actually," he said, "I'm meeting Sadie for tea."

These "tea-dates" with various women were planned in secrecy and kept secret until just beforehand. I was not included in the plans and not invited.

What the *hell*. He was sneaking out to see someone else, someone who had already been a problem in our relationship, and he wasn't going to tell me? It felt like whiplash, I had just been loving him so deeply and now I learned that he was being deceitful and cared more about meeting with another woman than staying in bed with me. That hurt.

And I am not blameless. In that moment when he said he was meeting Sadie for tea I should have made my boundaries clear: *her or me. This is the moment you decide.* But I didn't understand a thing about boundaries then. I knew the situation was crushing my heart, but that's all I knew.

At least I had the presence to say *something*.

"So, you weren't going to tell me? That doesn't feel right." I'm sure my voice waivered a little, although I was trying to stay calm.

And instead of staying and talking it out or deciding that a cup of tea wasn't worth destroying our relationship over, Jerry was moving toward the door, obviously late for his meeting time. And this is what he said:

"Well, this will be a good opportunity for you to process."

According to the way he thought, I was supposed to just breathe and put up with any sort of behavior, and it was my failing if I couldn't do that. It's like he was trying to train me to deal with disrespect. After he left I curled up on the bed and cried.

I thought I might give him one more chance, make it very, very, clear how I felt. Communication 101, at least I knew that much. "I am in pain." I texted.

He slowed the car for a minute, read the text, and *kept right on going!* He made a choice, and I did not come first. That was the moment he put a knife in our relationship. Sometimes I picture the years and years that came after with that spirit knife still dripping blood.

Now that I am a relationship expert (ha) I know that when difficult conflicts arise there are a few things that can help navigate through them. The first thing is to really listen to what the other person is saying and how much the issue means to them.

I could have done that for Jerry, too, I could have listened to why these tea dates were so important and ongoing – but he kept it all behind a wall, hidden from me. I learned to smell his deception and it pushed buttons I didn't even know I had. I got angry when I detected it, and I got better and better at detecting it. But Jerry was afraid of angry, he hid from it, and so we went around and around – not fun for anyone.

I could give you dozens and dozens of examples of when he said or did the 'wrong' thing and hurt me and damaged the relationship. But still I stuck it out, with my zero boundaries! Like a moth I kept trying, no matter how many times I got singed.

One time, as I was getting into his bed, I told him that the photo of another woman on his dresser bothered me. We were in an exclusive relationship at that point. Any other man would have gotten up, taken it down, and apologized. But not Jerry, mister 'stay the course' 'she'll have to learn to deal with it' 'I will not be manipulated.' Instead, he said, "that photo does not represent another woman, it represents a mood."

Well, to me it represents another woman and I think I'll go sleep in the other room now, I said without words.

That photo stayed there for months. He added one of me, as if that was supposed to make me feel better. Shall I go on?

Our first Christmas living together was coming up. Jerry was not really into 'things' so I thought deeply about what I might do for him. I decided I would make a website for his business. I had never done anything like that before, and I had to start from scratch, including purchasing the domain name, writing all the copy, and selecting all the photos. I put many hours into the project, and I was so excited to be able to present it to him. Guess what he got me? Nothing.

He wouldn't even take a hike with me because Sadie might be stopping by with cookies – she always brought cookies on the holidays.

The tea meetings continued, with a number of different women, and so did the deception and disrespect. I was in such a sorrowful place, and needed protection so badly, that I would literally look to the plants for it. The forests were a great comfort, but even mundane landscaping plants would help. I remember leaving a restaurant one day, with a heavy, heavy heart, and just standing, staring, and commiserating with the shrubs. The shrubs! *We are here, on this Earth, we are alive. We will be OK. Our circumstances are not ideal, but there is still beauty in the world.*

One day, in an effort to be completely clear about how I was feeling, I invited Jerry into my room looked him in the eyes and told him calmly, "I am very unhappy."

Now what would you say if your 'partner' told you that? I bet you wouldn't say what Jerry said. He said, "That's good! That's your ego dying."

We did have good times, but I suffered from the dark disrespect between them. I was still deep in my writing and forest-saving work, but the relationship turmoil was an unwelcome distraction.

We tried couples counseling. We went to three different people. But I spent my time there trying to get us closer as a couple and he spent his time there trying to get permission to visit his other female friends without getting me upset. I didn't like paying my half of *that* bill. To be clear, he wasn't having sex with any of these women, it was more emotional infidelity than anything else. He was after their *edges*, their secrets.

Reading these stories, you might think he was just not that into me, but that was not the case. One psychiatrist explains it this way: an unhealthy relationship with a parent can impair a person's ability to love freely and well. "His heart's gaze, in the manner of one whose eyes do not properly focus, will have the unsettling habit of looking beyond and behind the person in front of him. A heart thus displaced falters in its efforts to meet another's rhythms, to catch another's tempo and melody in the duet of love."[12]

Yup.

Eventually, I was able to look at my heartbreak from a distance: why this pain, and what can I do about it? I learned to welcome it. *Thank you, thank you, for this pain*, I said silently to the universe. It was showing me that something was wrong, and I must do something about it.

"Pain is a packet of chiseling tools," wrote Hoagland in his essay, *The Threshold and the Jolt of Pain.*[13] I was being chiseled alright, and the creature that came out of that time was different than the creature that went into it. Most people undergo heartbreak long before their fifties, but I had never before had to suffer it.

So brave, everyone had said. So brave after her husband died, so brave in leaving her job, so brave in moving to Colorado. But now I needed a different type of bravery – the bravery to give up. Pain was showing me the way. The incident that caused me to finally say "enough," was not even something that big. Once more he was having "tea" with another woman, once more I wasn't invited, and once more he had kept it hidden from me.

On a real estate website I spotted a little house for sale on a street I loved, named Grace. The house was back in Maryland, closer to my daughter, my friends, and my extended family. I made an offer without even seeing it. I would be moving across the country again, just a year and a half since my last move. This time my load was even lighter since I gifted some furniture, art, and rugs, and I gave away half my clothes and most of the electronics. I had successfully 'downsized' and, although it's a good feeling, I wouldn't wish the circumstances on anyone.

One day, after I moved back east, Jerry called and started telling me what a great couple we were, how great it would be if we could get back together, and how much we had in common. *So why'd you screw it up?* I was thinking. *Why didn't you even ask me to stay?*

All was calm and peaceful in my new home, except I was a little obsessed with recounting all the ways he had hurt me, well maybe more than a little.

I thought I had done well in healing from the grief of losing Rick, but right on top of that I was traumatized by my dysfunctional relationship with Jerry.

We talked on the phone often, as I thought that might be healing for me. I finally got to the place where I didn't hate his guts (as we used to describe those feelings in grade school). This was progress. He called me darling, told me he loved me, and he hoped that we could work toward being a couple again. I told him no way. The situation had flipped on its head but it was still not healthy.

Then I learned about childhood wounding. I was one of the many who thought I had a great upbringing and escaped with no damage. According to national polls, eighty-one percent of people say they had a happy childhood. But even if they think they have had a happy childhood, everyone has wounding. Many do not want to face their wounding because it feels traitorous to point out the ways we were failed by our parents, who were just doing their best. But for true personal growth we must face these wounds directly. It took me until the second half of my life to identify mine. My wound? I wanted to be special. And Rick made me feel that way, for decades. And that was my happy place, where I could blossom. Then, with Jerry, that all came crashing down and the wounds were deep. He treated me like I wasn't a priority. Sadie was more of a priority, or the cashier at the hardware store, or the next girl to walk into the restaurant with nipples showing...you get it. I began to wonder if wanting to be special was a failing, a weakness in me. All I knew was that I was very unhappy.

But my childhood wounding was nothing compared to Jerry's. While my father might not have noticed me, Jerry's father actively drank himself to death when Jerry was seven. Besides missing his dad, he was deeply ashamed to be growing up without one. Instead of focusing on her boys, Jerry's mother got busy trying to find another husband. Although Jerry was a star on the high school basketball team, his mother never went to a single one of his games.

Jerry's deepest wound was abandonment. He was always accusing me of having 'one foot out the door,' and it was true! Why would I want to put both feet into a relationship where I was not seen as special?

Toward the end he was really trying, he was more honest, more generous, and more focused on me, but even the *more* was not enough. There was something in his personality that was always looking for my weak spots.

Burning Man Three (age 63)

Why did I return to Burning Man in 2019? I was 63 years old – a definite outlier. In part it was because I finally found myself in a burner-friendly lifestyle. I was no longer teaching – so I didn't have to worry about the first week of school, AND I had a new Colorado friend with an RV that she took to Burning Man every year. My other two experiences had been in a poorly equipped van and in a tiny tent, and I'll admit that I was a little envious of the many – it seemed like the majority – who were staying in RVs. Now it would be my turn.

But those weren't the *real* reasons. In a life, even in a long life, we only have so many moments of *wow*. Only so many moments that are so marvelous or strange or weird or so rough that they will never be forgotten. Altogether I had spent less than two weeks at Burning Man, but in that short time I had gathered a bushel of *wows*. Meanwhile, whole years had been completely forgotten. I am a harvester of adventure and memory, those two sisters, and that was my real reason for returning.

But I was eight years older than I had been the last time, and I knew what to expect. In hindsight maybe I knew too much. I knew how easy it was to get dehydrated, to eat the wrong food, to become sleep deprived; or how easy it was to celebrate being there with too many drugs, or too much alcohol, or even how easy it was to get lost, or hurt. I had seen all of these things happen to others and I was determined they wouldn't happen to me.

When I was filling out the application form for our camp there was a space for my *Playa Name*. (Because you are a different person at Burning Man than you are in the default world, as burners call it, it is a tradition to adopt a playa name.) In the past I hadn't played that game, but suddenly it came to me: of course I had a playa name – it was *Dragonfly*.

In 2011, the day after my mystical experience with the dragonfly in the Temple, I was wandering through the outer ring of the camps in that afternoon hour when the light pours in like honey from low on the horizon. There was a man taking photographic portraits of burners in the golden light. His friend was nearby in a folding chair just relaxing and observing. I like watching photographers too, so I stopped and started chatting with the friend and for some reason my whole story came pouring out ... the death... the wedding clothes... the dragonfly. The look on his face told me he was moved by my story.

"Hey," he called to his photographer friend, "I want you to meet Joan, you've got to hear this." So I repeated my story for the photographer and he asked if he could photograph me. After the stories and the photos, when I was ready to move on, he gave me his card and said he'd be posting the photos online. A few weeks after I got home that year I looked up the photo site -- and there I was, titled *Dragonfly*.

I don't know why it took me eight years to realize that was my playa name. It is now also my Starbucks name. I love watching the clerk's eyes light up when I give my name, and I like the looks when my name is called as my latte is placed on the counter. *I never knew anyone named Dragonfly,* the other customers are thinking as they look my way. My friends, I didn't either, but here she is.

In 2019 the entry to camp went smoothly. We pulled in at one a.m. and knew exactly where we were supposed to park -- the camp had been mapped out ahead of time. There was none of the confusion Rick and I had experienced in 2009. After we parked I went right to sleep, but then woke at dawn, and slipped out into the fantastical world that is Burning Man.

I headed to Center Camp and the first thing that caught my eye was a huge painting of a dragonfly! I wanted a photo of myself with it but that wasn't possible selfie-style. I saw a rather plain looking guy walking toward me and I asked if he would kindly take my photo with the painting. He did, and we started talking. It's easy to make friends at Burning Man because as much as it is an experiment in built art, it is also an experiment in the art of community. People are open and helpful in a way they might not be in their home communities. Turns out that he had rolled into Burning Man last night also, but instead of going to sleep, like I did, he went out on his bike and had a grand time: bar camps, neon-colored art, pirate ships that rolled across the desert, topless girls, shared joints, music so loud you could feel the untz untz shaking your sternum, more bar camps, and on and on. Unbelievable!

But somewhere along the way he had gotten separated from his bicycle. That's the reason he was out now, at dawn, on foot, he was looking for his bike. I had to stifle a laugh when I heard his story. Unless you have been there you can't understand how futile a search this was. He had close to a zero percent chance of ever finding his bike again. It would be a long week on foot for him, and that would be *his* Burning Man story.

But this year I was traveling in style, and I had a real adult-sized bicycle strapped to the RV. When I got back from my walk my travel buddies were awake and we took our bikes down and decorated them with lights. Can you guess where I went on my first ride? The Temple, of course.

It was an unusual one this year, rather contemporary and minimalist: pairs of huge wooden towers, tallest in the center, forming a central shaded aisle. But the design doesn't really matter – it is the energy of all that feeling that truly creates the Temple. It is the signs and the memories brought from all over the world. There were photos of pets with their names written across them, I could feel the heartbreak from the loss of that special cat. I saw more than one small wooden box, presumably filled with ashes from a cremation, human or animal? Probably both were here. There were letters of apology from the recently divorced, and small delicate bouquets that may have been for the spirits of fetuses that never saw the light. On my first visit I was just taking it all in.

When I started planning my trip to Burning Man in 2019, the Temple was foremost in my mind. I was in a completely different head space than I was in 2011, but Rick's spirit was still very close to me, and I wanted to honor his continuing journey through the other realms. What should I bring to the Temple for him? Most of his material things were gone, and so were many of mine; however, I had kept some treasured textiles.

Rick spent time in Indonesia in the early 1970's, before I met him, and our first shared space was decorated with batiks he had collected while he was there. They seemed so exotic to me back then – I had never been out of the country! Eventually they became faded and worn, but I still didn't part with them. One of the batiks had an image of a chariot being pulled by horses, it seemed like a perfect tribute for the Temple, and, being the practical person I am, it was easy to pack. I took a marker and wrote a message to him across it. I packed the tacks again, too.

Meanwhile my mother, in her late-eighties, asked me to bring something to the Temple to burn. It was a small box of love letters her first husband, my father, had written to his lady-friend, and those she had written back -- while he was married to my mom! You might be wondering how these fell into her hands? Well Dad and his lady-friend ended up in a long and happy marriage and they kept the letters. As I already described, my stepmother died of cancer. A few years later dad died too. Their house was emptied and someone made the questionable decision to give the letters to mom. She didn't really want them, but for some reason she couldn't just toss them out. I was happy to help her release them. She said that I was free to read them, but after looking at one or two I decided that they were never meant for my eyes, so I closed the box for good.

On my second visit to the Temple that year I brought the batik and the letters. The walls were filling up fast. The letters found a perfect little corner – good riddance – and then I centered myself and found a spot for Rick's batik. Out came the tacks and my small hammer. *Rick. Oh Rick.* There was that old familiar swelling in my chest. *I love you. I miss you.* Then the tears. *It has been almost nine years. How can I have lived so long without you? How much longer must I wait to see you again?* By this point I was flat out weeping. I didn't care who was nearby, I didn't know any of these people.

When I finished hanging the batik and looked up I noticed a small group of people dressed head to toe in bright pink, fuzzy, costumes. One of them was walking calmly in my direction and the others seemed to coalesce toward me as well. The first to arrive extended his arms and I fell into them sobbing. It was a soft, warm, delicious embrace. The others arrived and joined in, and soon I was in the center of a mass of pink, hugging, loving, energy. There were no words spoken. My crying stopped. I wiped my face dry on their fur. After a few minutes we released our embrace and went our separate directions.

Now, it is possible that this was a group of friends who were camping together and dressed alike, and who just happened to be visiting the Temple at the same time I was. But I have a different story for them. I think that when they were deciding what gift they could bring to Burning Man, instead of a bar camp or an art car they decided that their gift would be comfort for the bereaved at the Temple. And how better to comfort than in a soft, colorful, gentle group. That is the story I believe, and I thank them.

Most of the time I hung out in my camp and worked my volunteer shifts at the Full Circle Tea House. That was the gift Camp Soft Landing (and its little sister, Camp Now) provided to the Burning Man community. The Tea House was inside a huge oval tent with tapestries and Moroccan lights hanging on the walls. The floor was covered in faded oriental carpets. Upon entering one felt like they had stepped into the enchanted domicile of a Bedouin prince. Along the walls of the tent, beneath the tapestries, were beautifully staged beds for tripping, or coming down, or sleeping. Couples in various stages of these activities usually occupied a few of the beds. In the center of the tent was a low, carved, wooden table that formed a circle with an eight-foot-wide space in the center, like a huge letter O. On both the inside and the outside of the table were red and purple cushions.

It was shady and cool (thanks to fans powered by solar panels) and it was open to receive guests at any time of the day or night. It was designed to be an oasis from the bright light, the heat, the noise, and the general craziness of Burning Man. When a person came into the tent they were greeted by someone sitting on a pillow in the center of the low, circular table, and they were invited to have a seat. That was my job. I was the tea lady.

I greeted my honored guests as if they were new friends, and asked if they would like some tea. If yes, their choices ranged from chamomile (the playa name of one of my camp mates), to puër, to a blend that included butterfly pea and rose. I channeled my inner geisha as I prepared tea on the tray between us. First, I measured the loose tea into the small clear pot, then the hot water, boiled to exactly 212 degrees, was poured from a metal thermos onto the tea leaves. I engaged my guests in conversation for a few minutes while the tea steeped. "How is your experience at Burning Man this year?" I asked. I shared about myself too, if they were interested. Then I poured the tea ceremonially into a tiny handleless teacup, and gestured for them to sip. As the cups were emptied I refilled them. I loved my role and I dressed for it with silk scarves, long jeweled earrings, and mala bead necklaces, over harem-worthy outfits. I was old(ish), but I could still feel enchanting. My dressing was not to attract anyone, it was for my own pleasure.

I was having such a positive experience at the tea house that I signed up for extra 'shifts.' I even volunteered to work the night of the Man burn, and I'm glad I did because that was the night I had my most memorable guest. I doubt she had a playa name, and I don't think I ever got her real name, but I'll call her Jesus Gal. She seemed to be in her late thirties, and she was wearing none of the extreme clothing (or lack of clothing) favored at

Burning Man. The tent was nearly empty that evening -- as almost everyone else was gathered to watch the burn.

So, Jesus Gal walks over to the low table behind which I am sitting on a pillow. I invite her to sit, and welcome her, as I had with many others that week. As I pour her tea she starts telling me her story. She had been at home, in a dirty New Jersey city, when she saw a news story about the huge Burning Man temporary community in Nevada. Likely there was footage of previous years' revelers to add the shock and awe component that television producers are so fond of.

"I need to be there," she thought. "I need to bring the Word of God to those lost sheep." She immediately bought a plane ticket, and with almost no preparations she was on a plane to the Reno airport. If you think Rick and I were unprepared in 2009... Jesus Gal had no way to get from the airport to Burning Man (a four-hour drive), no ticket to get in once she got there, and no shelter, not even a sleeping bag, if she did make it.

Yet here she was in front of me sipping tea. The long story was filled with miracles and wonders. She found a ride from the Reno airport, without even posting or asking, and then she stayed at another stranger's house who insisted she couldn't go to Burning Man without a coat or a sleeping bag (they gave her both, knowing they'd never see them again). The details are a little fuzzy now, but she was offered a ride into the event. She explained to the driver that she didn't have a ticket. (The $425 tickets had sold out five months ago, just minutes after they went on sale.) When the car got to the gate – where vehicles are thoroughly inspected to make sure no one is hiding—she walked up to the 'will-call' booth and explained that she had no ticket but she felt called to be there. The person in the booth looked incredulous, and then pointed her in the direction of two men sitting on the ground nearby. Black market tickets were selling for a thousand dollars each, but the men just handed over their extra ticket for free. She was in.

And then I was serving her tea. I don't know where she camped or what her experience ended up being, but I like to think that she was surprised by the spiritual nature of many of the participants – something the evening news doesn't cover. I got the feeling she was sharing her story so I might suddenly see the light and fall to my knees like apostle Paul on his way to Damascus. But, darling Gal, don't you see? I am already on my knees, serving you tea. You don't need to save me; I know Christ Consciousness. Miracles and wonders are nothing new to me.

That year I knew exactly how I was getting home and I had finished all my volunteer shifts. On our last day my girlfriends and I could relax together. Around mid-day we hopped on our bikes and cut out across the hot, dusty desert. Our destination was something like a human car wash. When we arrived we got on line, but there was a rumor that they would be closing down the human-wash camp soon. They were running out of water, and anyway, the week was just about over and dismantling of camps was about to begin. Those of us in line danced to the music and begged them to stay open for us. Occasionally a camp volunteer would count heads as they walked down the line. How many more could they take?? We all wanted to know.

We breathed a sigh of relief as we entered the place where we were formally reminded of 'appropriate gaze,' this meant we were going to make it in. Eventually it was our turn and we stepped through the canvas door into a room of about one hundred-fifty wet and naked humans dancing their asses off. *Well, hello.*

Our group of ten were the only clothed ones, but that didn't last long. We were instructed to take our clothes off and leave them in the one of the cubbies. Like good children, we did as instructed. When all were finished with that task, we were led into something that looked like a giant plastic gerbil nest. Across the top of the tank were walkways, and standing on the walkways were gorgeous young women who hoisted giant hoses that sprayed warm soapy water, of the perfect temperature and pressure, down onto our dusty, naked, bodies. They were dancing, and laughing, and making us turn this way and that while they hosed us down. And we were dancing too. Oh joy!

My friend told me the soap was Dr. Bronner's, and the camp was assembled and paid for by that company, but there were no signs, and, shhh, you shouldn't even talk about it, because there is no advertising or commodification allowed at Burning Man. This human wash was a hugely generous gift.

Next, we were ushered into the rinsing tank, where we were hosed down again, and at the end of that we were guided out onto the dance floor. Now *we* were the laughing, dancing, clean, naked people being witnessed by the last and final group coming in. And if all this were not enough, as I was dancing I saw that there on a narrow stage, on the side of the room, was the famous artist Alex Grey, working on a huge painting of this very scene! I adore Alex's paintings so this was a treat for me. His wife, Allyson, was there too. I had met both of them before, at their home in upstate New York, so I went up and we had a little chat, but the bumping music was quite loud so we kept it short. I went back to dancing. The energy was so, so, high in that room! It was the last washes this year, and we were all so happy we made it, and so clean, and so ready for all the lessons to sink in and ready for the future too.

Alex could feel the energy zoom through the room. This was *it*. He pulled off his clothes and jumped down from the stage and the next thing you know I was dancing with famous artist Alex Grey, both of us completely naked.

When it was all over, and we were dressed and ready to go, my girlfriends and I stepped outside into the bright world feeling deliciously clean. A man who called himself 'Boy Scout' rode up on an adult sized tricycle pulling an insulated metal box on wheels. "Would you ladies like a glass of cold white wine?" *Well, yes, that sounds marvelous, thank you.*

Looking back on all three Burning Man experiences, the first year I was with a partner who could barely walk, the second time I was a grieving widow, and the third time I was in my sixties and way too careful with myself. I don't think there is such a thing as a typical Burning Man experience, but mine may have been especially atypical.

Each time, being separated, for a week, from plants, and especially trees, was difficult but it allowed me to see so clearly the generosity and creativity of the remarkable animals we call humans. Imagine if another animal dressed up in different outfits each day, built amazing structures, and interacted in crazy ways. We would be paying great sums to line up just to watch them. But here we are, they are us!

Zipolite Three

As Jerry and I struggled about everything else, we struggled over Zipolite too. At first I wanted him to go with me, but he didn't want to. Then he wanted to go, but I didn't want him to. This awkward dance went on for years. Then came 2020, the year I really wanted him to go, and he, too, really wanted to go. The plane tickets had been purchased and I had reserved the best suite on the beach for us: a place of first-class architecture, complete with folding glass doors that opened onto a terrace overlooking the ocean.

Although I lived back east by then, I flew to Colorado so I could accompany him on the journey. The day before we were leaving, as we checked in online for our flight, I discovered that his passport had expired. Because it was mid-pandemic no passport offices were open for expediting. There was nothing he could do. Sorry babe, my passport is good, that beautiful room is waiting for me, and I am no longer a people pleaser.

We had made a plan to take a side trip (wink) while we were in Mexico together. We each packed a chocolate bar that was infused with magic mushrooms, made for us by a super-cool California friend. When it became evident that I was traveling alone, again, we decided that we would still trip together, but over Zoom.

When the appointed day and hour arrived, I settled into a comfortable spot on my terrace with a vast view of both the ocean and my beloved rocky headlands. I opened my little Mac laptop and logged on, there was Jerry on the screen. We ate our chocolate bars together while setting intentions for our journey. As the ingredients made their way through my bloodstream and into my brain the world became more and more beautiful. I could look past my computer screen to see the shiny green palm leaves gently swaying, and behind them the blue-blue ocean with white-capped waves. Over the waves swooped formations of pelicans, and, closer in, orioles flitted amongst the colorful bougainvillea flowers. Now and then a hummingbird would check out the flowers on my Hawaiian print dress, or a tiny gecko would scurry up the terrace wall. I was living in an utterly, utterly, beautiful magical world!

But it is always thus, of course.

In the foreground, I could see my computer screen with two images on Zoom, one of them was me, reclining on the terrace with my blue-eyes alight with the beauty of the scene around me. Such a lovely woman, my beauty matched the beauty of the world.

Jerry and I weren't conversing much, maybe just some astonished laughter about how freaking perfect and alive and vibrant everything was. After an hour or two of that I felt an urge to stop being witnessed, and to stop witnessing, and to go deeper on my own. With a brief explanation I closed the computer. Then I closed my eyes. That same life energy I was witnessing all around me could now be felt inside of me. And the interior was just as beautiful as the exterior.

After a brief meditation I lit a stick of copal incense. As I watched the blue-grey smoke rise I felt that it was in perfect sync with my mind. A curl of my thought patterns produced a curl in the smoke (or was it the other way around?). The loops and curls of my mind and the smoke were moving together and completely connected, yet ever-changing, like the murmuration of a starling flock.

Collective behavior in both living and non-living systems has been observed before. For instance, when atomic particles are placed in a large external magnetic field, their spins will align. The smoke from the copal and the electrical wave impulses from my brain were all in the same magnetic field, so perhaps we were entrained in some way beyond mere hallucination. Also, our electrons were in proximity, and therefore interacting with each other; physicists have termed this *hyperfine interaction*. When neurons fire they create a disturbance in the electromagnetic field. Who is to say that my psilocybin influenced brain neurons were not affecting the electromagnetic field of the smoke, and thus its patterns?

One researcher believes that the electromagnetic field comprises a universal consciousness that experiences the sensations, perceptions, thoughts and emotions of every conscious being in the universe.[14] You may discount the relationship between the smoke patterns and my brain waves, but that experiment has not been done. Whatever the causal agent, the synchrony between consciousness and smoldering copal was mesmerizing.

And this was not just any incense, copal[15] has been a sacred substance since before recorded history. Even ancient Aztec temples contain representations of copal use. In present day it is still associated with the spirit, with the holy. I have been to churches in Chiapas, Mexico, where I couldn't see the altar because there was so much copal being burned.

Historically, the white smoke produced by burning copal resin was associated with helpful spirits and "White Gods." The smoke was considered food for the Gods, and it assisted in invoking deities and helpful spirits who were nourished by the smoke and, in turn, responded to human pleas.

If this copal smoke experience sounds far-out, wait until you hear what came next. After an hour or so of copal entrainment I looked up into the clear blue sky and saw the "Face of God." It resembled no human face, and if I had to put a description to it, it was more like a pattern of circuits in a circuit board, with various colors lighting up the nodes and lines. It reminded me of the artwork of the Shipibo people of Peru. It was beautiful, and captivating, and I was not afraid. We communicated, but I cannot relate any specific messages. The communication was more of a *feeling*. I was connected directly to the energy of God and I knew it. I wanted to hold on to it as long as I could. To anyone witnessing this scene it would only appear as if I were calmly staring into the sky, but on the inside I was in awe, completely filled with the spirit, completely loved and loving.

This was another touchstone I knew I would remember forever. As I age it seems that God and I are getting closer and closer.

It lasted perhaps thirty minutes. When the vision faded, I moved to my bed where I could process what had just happened. The sun was setting and my room was shady and cool. As I laid in that big beautiful space I felt that I was ready to let go of life. Truly, I would have been completely satisfied with dying in that moment. I prepared myself for letting go of any earthly reins. Just this small body, ready to release its spirit. So easy, so welcome. I laid there 'dying' until the room got dark.

Until...until thoughts entered my head of who my leaving would affect. My daughter! My grandson! My mother! Oh, we were so connected! It felt like my heart was being squeezed. They needed me so much! I felt the connections deeper than I ever had before. We are all part of the 'great weave.' If I left my body now it would be wrenching to them, especially my mother. Her husband of over fifty years, my stepfather, was at home in hospice care. Suddenly, like a bolt of electricity, I knew that he was actively dying and she needed me. I should go home immediately.

There are no night flights from Zipolite, and I was coming down from a mushroom trip, so I didn't start throwing things in my suitcase and running to catch a taxi to the airport, like I would have in a Hollywood movie; instead, I had a meal of rice and beans and went to sleep.

In the morning I called the airlines and arranged for the next flight home.

I went to visit my mom and step-dad as soon as I got there. It was his last day alive. I think he knew I was there. More importantly, I was there for my mom, because God had shown me how deeply connected we are, each to the other, and how we are more to each other than we can ever say.

That "Face of God" experience was in 2020, during the height of pandemic hysteria. It was bizarre walking through airports with none of the restaurants or shops opened. Everyone eyed each other as a contagious enemy. The flight attendants handed out sanitizing wipes as we came aboard. No food or drinks were served.

In 2021 the pandemic was still a thing, and I knew some people who would have criticized me for not staying safely at home. I didn't post my travels on social media those two years. Why stir the hornets' nest? But I knew that I needed to return to Zipolite, for both my mental and physical health. In 2021 I decided I really preferred to go alone, sorry Jerry.

That was the year I met Julian. I was loving my solo retreat in my perfect little cabana on the beach. I had a routine that involved coffee, the sunrise, a yoga class, a swim, fresh squeezed juice, a marvelous meal, and a shower, interspersed with lots of reading, walking, and sky watching. One morning as I walked out of my yoga class and was slipping my sandals on, I noticed a very handsome man with long blonde hair, blue eyes, and strong dark eyebrows. He was alone and I did something uncharacteristic for me – I glanced at his finger to see if there was a ring.

There was none.

I struck up a conversation, also uncharacteristic of me. I was not looking for a boyfriend, or even a friend, I was happy in my little introverted world, but this guy seemed interested in me too. He was an American traveling alone, and he would be leaving for Oaxaca City later that day.

As part of my usual routine, after yoga I always went for a swim. While we talked, he walked down to the water's edge with me, and when we stopped I started taking off my clothes, every last scrap of them, because that's how I swim in Zipolite. I didn't even have a suit with me.

I don't think of myself as an exhibitionist or even as a naturist, I just find it highly practical to swim, dry off, and then get dressed, without a wet bathing suit to deal with. This was routine for me by now, and I didn't think much of it.

What's a guy to do? He started stripping down too. So there we were, naked in the clear aqua water watching the little minnows swim around, having a nice conversation, before I ever learned his name. And just as I had noticed that he had no ring and the bluest eyes, I also noticed his nice body and the crystal hanging around his neck.

After we dried off, dressed, exchanged contact information, and went our separate ways, we bumped into each other again, an hour later, in town. Over a coffee I told him all about Peter Pan, how Peter had died that year and my plan for the afternoon was to visit his grave. Julian offered to give me a ride to the town cemetery. Together we wandered through the raised concrete rectangles looking for Peter's resting place. Quite an unusual first date!

The following year I returned to Zipolite alone again. I saved money by renting a rustic casita off the beach. One morning at the beachfront restaurant, while waiting for my juice to arrive, I recalled being in that very spot in my forties, decades ago, and watching a nude, elderly woman, wade into the waves. I found her interesting to watch because she seemed so frail. It was easy to tell her age from a distance because of the way her white skin sagged off her bones. Back then I wondered what the old woman was doing there alone and unafraid of the strong surf. In that moment of memory, I realized that I am now just like that woman, and perhaps the exact age she was when I was watching her. The mobius strip of time had twisted once again.

Time continues to peak and collapse, wavespell upon wavespell. As this story continues I become more and more aware of my aging. I am no longer wishing my life away, but I become more ready to move on from this particular human form.

It's likely that others now watch me enter the water with a similar curiosity. I don't care, and probably she didn't either. Although I have stayed healthy, our bodies tell the story. My formerly perfect ass (according to more than one admirer) is now drooping. My small breasts have lost a little, shall we say, *inflation*. My arms and legs are strong, but the skin is blotchy. My hair is gray. I am unembarrassed by all of this, I just want to submerge in the salty water, I want to welcome the energy of Mother Ocean. As I enter her, I pause for a moment of gratitude, "Thank you for my body, thank you for this beautiful Earth, thank you for allowing me to return here again."

The day after I recalled the old woman in the surf, I was standing up to my waist in ocean water the color and clarity of old-fashioned coke bottles; nearby was a handsome young man – at least two decades younger than me. His hair was medium-brown and trimmed close to his head. His body was toned and tan, his cheeks rosy-pink, like a cherub, I thought. He seemed the personification of 'sun-kissed.' The pinkness in his checks set off his turquoise eyes, but otherwise he was a warm, pale brown, over his entire nude body. We were the only two people on this secluded little beach beneath a place called, no kidding, *Shambala*.

My body was nude too, but, as I just described, very white and very sixty-six. I had no idea what his name was, nor he mine, but we were enjoying a nice chat. He was Canadian. Turns out we were both yoga practitioners – in different studios. Each of us had already completed a rigorous yoga session that morning, and now it was ten a.m. – time for a swim.

By this point we were standing very close to each other in the challenging surf. He looked me square in the eyes and I looked back, unembarrassed. "A few minutes ago you told me your age," he said, never losing eye contact. "I think you are beautiful...look," he said, with a slight downward gesture, "my body is responding to you." I glanced down and, sure enough, he had an erection.

With all this talk of nudity and men you may be wondering which direction this story will go now, and for a moment I was wondering that myself, but quickly my flee response kicked in, and I made some excuse about needing to get out of the sun.

Was the flee response coming from my trauma with Jerry, my spiritual growth, or just plain common sense winning out over sexual urges -- at last?

I made my way back up to my rustic cottage on the hillside. I sprawled across the king-sized bed and the answer came to me. It was *space*.

My friends want me to find love again. Every trip I go on they think, this is it, this is the time it will happen. But again and again I return, still single. They think I'm trying to find love, *should* find love, but I at this point I am loving the space, both in my daily life and in my heart and soul. I would rather have this space than a brain full of memories and thoughts about what *might* have occurred that day. I recognize that this attitude is a part of aging, for better or for worse, and I welcome it. I am my own priority now, no one else has to cast me in that light.

More than ever, this beach is a retreat for me. I rise with the sun, meditate, walk to the yoga studio, get real in my skin, go for a swim, look at the trees, look at the clouds, look at the trees, look at the birds, look at the interesting bodies. We all have one. It is not even noon. Time for a shower and a sway in the hammock. Maybe some writing after that. Yoga is union of body and mind.

Bhutan on my Mind

Was there no end to the wild things I would do? I had signed up to walk the entire width of a small mountainous country far, far, away from my home. How had that happened?

In 2019 my sister said, "Let's go on a trek together while we still can."

"Yes, let's!" I answered enthusiastically. I knew of no one else with the physical and financial resources, not to mention the desire, who was willing to do something like that with me.

"Where shall we go?" she asked.

"Bhutan," I answered without hesitation, surprising even myself. I had always wanted to trek in the Himalayas and Bhutan seemed the most far-out place I could think of. Even Rick, with his bad knee, wouldn't have been able to trek there with me.

After a little research we found a great ten-day village to village trek that we were going to book in spring 2020 for the following fall. But we all know what happened in March 2020, when the world's population hunkered down and masked up. Bhutan, like many other countries was completely closed to visitors, and we wouldn't have risked going then even if it wasn't. So much had changed by the time the country reopened in fall of 2022. For one thing, during the years of the closure, many of the people normally helping tourists, were instead working on repairing and reestablishing a trail that went across the entire country – the Trans Bhutan Trail, TBT for short.

The PR for the new trail was captivating: *After a 60-year hiatus, the Trans Bhutan Trail is set to finally reopen. Since the 16th century, when the 250-mile trek was the only way to travel across Bhutan, the trail has served as a pilgrimage route for Buddhists.... Now it has been reimagined as an outdoor adventure through the world's first carbon-negative country. Intrepid travelers can hike, bike, and camp through the lush meadows and dense forests.... Completion of the trail required the restoration of 18 major bridges, 10,000 stairs, and hundreds of miles of pathways. The route is peppered with museums and ancient fortresses...*

Soon I was watching webinars about the new trail and imagining myself on that trek instead of the modest ten-day journey we had formerly chosen. By going in the winter 'off season' when fees were cheaper, the price came to a figure I could see myself paying. It was, by far, more than I had ever paid for any journey, but I felt ready for one more, likely last, big adventure.

My body was strong and fit, but let's face it, I was sixty-something and I had only to look in the mirror to realize that I was slowly on the way out. Among other changes, my neck looked nothing like it did ten years ago. If I was ever going to do a big hike, now was the time (or never). It felt like a current was pulling me toward Bhutan, I was like a particle headed for a black hole. I still had my doubts, but I noticed myself buying thick new hiking socks.

I made a phone call to my sister. An expanded trip like this would be just mine. She understood. Very quickly the trip fell into place. I would leave in a few months. The TBT was just opening and I was to be one of the first foreigners on it. And that is how it became a solo trek, with a guide.

Why in the world would I want to do this? Why indeed. Some sort of ego trip? Some desire to push myself up mountainsides one last time? My risk-taking personality showing itself again? A desire to be deep in nature and away from all my electronic devices for a month? A nice alliterative title? Why does the word *fiercely* get so often paired with *independent*? I wouldn't call myself fierce. So, what then? Strongly, confidently, unapologetically, independent? Yes, that's more like it. Finally, I accept that it is just my dharma to always be pushing it, to always be on the edge.

Some trees are also like this, the 'pioneer species' that are the first to grow in a cleared and then abandoned field. They stand alone above the smaller weedy species. Pine, I see you, cherry, I see you. These first trees slow the wind, give birds a place to land, they drop their leaves and needles to make the soil beneath them richer. Their roots form relationships with the below-ground fungi. Eventually they create a safe space for the less hardy tree species.

It looked like I was really going to do the big trek, and a visit to the Tiger's Nest monastery would be the cherry on top.

The first time I saw a photograph of the Tiger's Nest monastery in Bhutan, the image was burned into my brain. After that, I could see a photo of it anywhere and I knew exactly what and where it was.

The dentist I went to for decades liked to take exotic vacations and practice his photography. In his waiting room were images from around the world: the pyramids in Giza, the canals of Venice, but it was the image of the Tiger's Nest that captured me as I sat there waiting my dreadful turn. He had <u>actually</u> been there! My esteem for him was instantly magnified. People can go there, and he had gone there. Oh, how I would love to go there myself, I swooned. But it seemed so, foreign, so remote. I had plenty of adventure in my life, but my imagination never stretched all the way to the possibility of really visiting Bhutan. Until now.

When I started sharing my plans with friends, almost everyone had the same question: "Where's Bhutan?" My shorthand answer was, "it's between China and India, and next to Nepal." The *where* was one thing, but the *why* was something else. As the poet William Stafford writes, "People wonder about what you are pursuing./ You have to explain about the thread."

It amazed me that this Buddhist nation, that had held such a place of significance in my mind, had never even crossed the minds of the majority of people. In this way it was like Burning Man, something else that was on my mind, and seemingly no one else's in the 90s. Besides the call of the Tiger's Nest, why Bhutan? Maybe it was because of descriptions such as this one I found in a book by a young American woman who went to teach there:

We turn off the main trails, following narrower tracks into forests, through fields. I am no longer dismayed at the way a wide, worn trail can splinter into a dozen smaller paths, one of which winds down a slope and disappears at a log. We climb over the log, slosh across a stream and another path picks us up, carries us through rice paddies, to someone's backdoor. A dog chases us around the kitchen garden into the forest, where a path brings us up to the road. There are always large stones to sit and rest on, trees to sit and rest under, there is no restricted place, no lines and bars separating what clearly belongs to someone from what belongs to everyone.[16]

I was hungry for a world without lines – even if it was only in a small corner of the globe where that could still be found.

I was born in 1956, so I remember the 1960s. I remember the night the Beatles played on the Ed Sullivan show, I remember watching Vietnam war scenes on the television news, and I remember when we landed men on the moon. That's what life was like in the United States; meanwhile, in Bhutan, in the year I was born, there were no roads, no telephones, no postal services, no airports, no hospitals, no national currency, and no television. [17] Today Bhutan has all of those things, but it is still one of the least-developed, least populated, countries in the world. My small state, Maryland, has a population of about six million people; but the entire nation of Bhutan has only one-eighth that number of people.

The first foreign tourists were allowed in Bhutan the year I graduated from high school, and they were allowed only under very controlled conditions. Until 2022 all tourists whether in groups or individuals, had to travel on a planned, prepaid, guided package-tours.

Bhutan is the last Buddhist kingdom in the Himalayas. Nepal fell to India, Tibet fell to China, but little Bhutan remains independent. It is ruled equally by the king and the monks.

I spent hours each day thinking about the trek. Often, I imagined myself gliding through the countryside with a smile on my face and a song in my heart. What a nice image! But I know that I tend to be an optimist. In other, perhaps more realistic, moments I imagined myself in Bhutan panting up a steep hill step by step, perhaps in the rain, perhaps with a sore ankle, and wondering what in the world I had gotten myself into.

There is excitement around the trip, but there is trepidation also, and as frequently as I imagine roaming past tall trees and soaring peaks, I imagine the abscessed tooth that will flare up when I am far from a western dentist, or the sleepless nights spent cold and shivering. I take another look at the itinerary. What once seemed very doable now looks rather frightening. Day after day of hikes over ten miles up and down high-altitude mountainsides. What was I thinking? *It's not too late to change your mind*, I keep reminding myself.

Finally, I settle in to an empty-mind acceptance. As Van Morrison sings in his song *Summertime in England*: "It ain't why, why, why. It just is." Sometimes things call YOU and not the other way around. In one of Rumi's poems he explains that there are those who are drawn to the water and those who are drawn to the fire, each thinking that God's presence was in the element they chose. In my mind that pair of opposites could be anything. In this case they could be safety and risk. (Why risk a trip like that, why not stay safe at home?). But Rumi explains that no matter which of the opposites you choose you will wind up on the other side. So staying home I could fall down my steps and end up with a major injury, and on a "risky" journey I could come home healthier than ever. As Rumi concludes, rather mysteriously: "The fire and water themselves: accidental, done with mirrors."[18]

I have never been a gym rat, pretty much don't like the gym experience at all, but I have signed up for a month-long membership and promise myself I'll go every day. The StairMaster becomes my beast of burden, turning worry into action. The first day I lasted about a minute before my heart started pounding. By the end of the week I could relax enough to start reading the signs on the walls: *If it doesn't challenge you it doesn't change you.* At first this mantra was encouraging, but by the end of week I was reading it a different way. What if challenging myself changes me in ways I don't want to be changed?

My biggest fear is that I would leave for the trip feeling fit and free of any aches and pains (my usual blessed condition) and I would return having done chronic damage to my body, causing me to live the rest of my life in pain. Would that be worth this adventure? I think not! So should I not go? Should I let this fear stop me? I honestly don't know, so I just move ahead. I must believe that something has propelled me, and when I'm fit and pain-free is exactly when I should go.

But those were also the thoughts of ski-mountaineer Hilaree Nelson. She was named one of the 25 Most Adventurous Women of the Past 25 Years[19], and was a member of the Explorer's Club, like me. But now she is dead, at age 49, after being caught by a small avalanche and falling almost 6,000 ft while skiing down a mountain in Nepal. I know I can't really compare my travels to hers; but what challenged her certainly changed her. She is now a pile of ashes, and her teenage boys are bereft.

When I told another elderly friend that I was going to the gym every day to prepare for my trip he said, "Well, it will either build you up or tear you down." Wise words. As we age, and increase our physical activity, we like to think we are building our bones and muscles, but maybe not. My friends who are runners have needed more surgeries than my friends who are not runners. But being careful with our bodies doesn't work either; one woman I know, in her eighties, was careful with herself. Recently she was at home playing cards with a few friends, and when she stood up at the end of the game her spine cracked -- and bone protruded through her skin. An ambulance was called. So, I guess it doesn't matter what we do, although I was working to stay in shape.

One morning, a few weeks before the trip, I woke with 'the fear' and carried it with me through the day. *It's not too late to change your mind,* I told myself. Then in the early evening I saw this post from a friend, a friend who is nursing her husband who is paralyzed from a fall that almost killed him:

```
if you give up. or do nothing.
you know exactly what will
happen. the outcome is simple.
so life becomes simple. nothing
changes. and neither do you.

that can feel comforting. but
if you really ask yourself
what you want out of life it
wouldn't be "nothing." it would
be "everything."

so go forward. do those things
that make you nervous. that
make you question everything
all at once. because we often
find out where we are meant to
be when we go to places that we
were once afraid to go.

/ topher kearby
```

I have been a student of the Vietnamese Monk, Thich Nhat Hahn, for years and he puts it this way, "The miracle is not to walk on water. The miracle is to walk on the green earth, dwelling deeply in the present moment and feeling truly alive."

So, finally, I say, f*** *yes*. I am going. Whatever happens, happens. If I can't complete the trek, well then, I will have some other amazing experience. I have to trust that the thing that is meant to be will happen. And so be it, I am going – into and through the looking glass.

Just a few days to go now. I am preparing as if I will never return, and who knows? I may not. I remind my daughter that my end-of-life paperwork, and my online passwords are all in the safe deposit box. I spend a delightful evening drinking wine with friends, and even as we are laughing I am aware that this may be the last night we share together. Instead of this being sad, it makes our time together that much more precious.

For the first time I lose sleep over the trip. *WHY* am I doing this?? I'm not sure I can answer. Why have I done anything I have ever done? Why did I get a PhD (leaving my family for long periods of time)? Why did I start a non-profit (risking poverty and/or failure)? But these just seem part of my unfolding story, so I reassure myself that this trip is too.

Bhutan for Real

Entering the airport in Bhutan to go through customs is like walking into a dream. There is a fabulous skylight illuminating an open space surrounded by wooden walls covered in intricate paintings. There are no lines, no advertisements. A young man wearing his native clothing, which resembles a tailored plaid woolen robe and knee socks, motions politely for you to step up to the counter. The counter is made of polished wood painted with the same intricate designs as the walls. In less than five minutes your passport is stamped, and you are heading out the door to collect your luggage, already waiting there for you. My experience going through customs in the magical airport reassured me, but when I got to the baggage conveyor and lifted my duffel bag onto the cart my back 'went out.'

It was like a spring had suddenly unwound and pain was released. If you are like the majority of the population, you have had that experience. *But not in Bhutan when you are leaving on a month-long trek!*

"Oh _now_, really?" I asked myself and the universe in tandem. I hadn't had any back pain at all in four decades, and *now* my back was going to go out? I was dejected as I shuffled out the airport doors to meet my guide.

The very first thing I saw when I exited the doors was a dragonfly! What great relief that little creature brought me. I felt like I was being watched over by a higher power. I was here now, I reminded myself, and I was blessed to be here, and whatever would be, would be.

I have always wanted to be one of those people greeted with their name on a sign at the airport, though it had never quite worked out. But right there I saw a middle-aged gentleman holding a sign saying "Joan Maloof." It was my guide. He took my bags off the cart and carried them into a car and presented me with a white silk scarf – a traditional gift of welcome in this Buddhist culture. Ordinarily this would have been a very special moment, but I was distracted. He asked what I needed: water, food, restroom? Well, I told him, what I needed was a flat spot to try to work the kink out of my back.

I doubt I was the trekker he was hoping for. *Look at this fragile old lady I will have to care for* (this is what I imagined he was thinking). We looked around and saw a stone shelf lining the parking area; that would have to do. Within minutes of my arrival in Bhutan I was on my back doing yoga stretches in a parking lot full of people. Not at all the entry I was expecting. I was hoping to have at least a few days before my limitations became apparent.

Our first stop was a stupa built for the 4th King of Bhutan who, while on a safari in Africa, died of sudden heart failure at age 44. My guide told me the tradition was to walk three times around the stupa in a clockwise direction. As I walked I smelled the incense that would soon become so familiar, I saw the rows of butter lamps burning, the colorful prayer flags waving, the elderly spinning their hand-held prayer wheels, a young woman prostrating herself over and over. My first two times around I was just taking in all this strangeness, but on the third time my eyes were streaming with tears that I couldn't attribute to a single emotion – I was feeling them all.

That first night I stayed in a nice hotel in the capital – the smallest national capital in the world -- but I didn't sleep well, between the back pain, the stretches I was doing in hopes of making it go away, and the mind chatter.

I was thinking of my friend who is a chiropractor; when he introduces himself, he says that he specializes in *skeletal-muscular pain.* I am always half-jokingly bragging to him that I have no skeletal-muscular pain, whatsoever. Oops. I also thought of Rick whose back went out often, and his failing knees, and how I was always urging him to do things out of his comfort zone. Now I could relate.

Then I thought of my friend Sam, who had such back pain it was ruining his life. After he checked in to Dr. Sarno's clinic everything turned around. Sarno claims that the back pain is nothing physical, instead it is rooted in our psychology. We shouldn't believe the pain is real and let it try to control us. When our back hurts we shouldn't baby it, instead we should be asking what it is trying to tell us. I'm afraid I knew what this pain was telling me: *you are out of your league, you are a chicken shit, and you should go home.* Or was the message, *time for you to learn a little humility*?

Then I remembered Rick's journey with the Sarno method, he tried his best to ignore the pain -- until he learned that his spine was crumbling from cancer. After he died, I couldn't bear to look at Sarno's book, and I threw it away.[20]

On my second night in Bhutan I already I want to leave. It was not because of the present conditions. Outside of my window, at sunset, I watched two young boys thresh rice by hand as ruddy shell ducks flew in small flocks above them. I was staying in a charming room, in a beautiful traditional building, next to a wide, clear river, running through a picturesque valley. My distress was because my guide and I had just looked at the topographical maps for the first time. (He had not done the TBT yet either.) It was going to be a grueling hike! I was sure I'd die of a heart attack or stumble and break a bone. *Hope you enjoy being medevac'd out*, I told myself. Why, oh why, did I ever think I could do this? What is wrong with me? I tossed and turned in my bed trying to sleep.

I am old enough to know that these midnight crises are self-inflicted. I tried to talk myself down. Nietzsche said we should learn to love our fate, so whatever happens we should say, "this is what I need." And when the worst things happen we should be grateful, because those are exactly the circumstances we need to help us grow, and in that way there is nothing that is not positive. Just think of the lessons I will learn if I have a heart attack on the trail! Will I someday think that is the best thing to ever happen to me? And what if I die, so what? I don't care personally, but the thought of my grandson grieving my loss is a tough one to let go of.

So, I did what I usually do when I face a challenge – collect more data. I got up and turned on the lights, got paper and pencil and started looking carefully at the maps. Now which hikes, exactly, were too much. Or how many long hikes back-to back would I really want to do? Three weeks of daily ten-mile hikes was too much even for the younger me. I noted hikes that seemed too long and others that seemed just right. Then I went back to sleep.

At breakfast the next morning I said to my guide, "we need to talk." I told him I didn't really care about how many Ks (Kilometers) I checked off, but I wanted to enjoy the trek – likely my last long trek -- instead of dreading it, and I wanted to go home healthy and whole. With that in mind, I thought we should adjust. He was completely agreeable! And what seemed so worrisome the night before evaporated in the morning with the clouds that were burning off from the hillsides and revealing blue sky. Time to head out!

And just like that, it was a brand-new day. We started by visiting the most famous building in the nation besides the Tiger's Nest. Everything about the Pokahara blew my mind from the arched wooden foot bridge covered with intricate painted designs (yes, the same paintings all over the country), to the fish swimming in the clear cold river below (protected because they are near a temple) to the active honeycombs draping from the high ornate windows ('spiritual' evidence of wealth contained within -- because honeybees are attracted to wealth). But it was within the temple, where shoes must be off and no photographs are allowed, that the real magic happened. Here all national ceremonies from coronations to marriages of the kings take place beneath the watchful, though nonjudgmental, eyes of many large golden buddhas. The surrounding walls are covered with intricate paintings of the Buddha's life. On the floor are velvet maroon cushions where meditators can sit. The small collection of visitors included monks, Japanese tourists, and Bhutanese families. I was the only 'westerner'.

I found my own cushion out of the way and sat. I had already sat in meditation in my beautiful hotel room that morning, but this was something completely different! The air was electric and I dropped in quickly and deeply. The space behind my closed eyes was alive and waves of currents passed over my skin creating what I call goosebumps. When that cycle was complete I opened my eyes and there was the golden Buddha of the Future looking at me. We fell in love.

Lest you get the impression that I am some sort of spiritually advanced holy person, one of my first thoughts was "I wonder where I could buy a smaller version of that statue." And this was not a unique thought, a version of it had been occurring all through my trip here thus far. When I was at the festival and admiring the lovely female dancers, part of my mind was thinking: *If you were going to purchase one of those kiras (dresses) which one would it be? And where would you wear it? And those clown costumes would be great for Halloween. I wonder if I could purchase one?* Etcetera, etcetera, etfuckingcetera! The funny thing about this is that I am not really a shopper or spender at all. In fact, so far on this trip to India and Bhutan I had purchased exactly – nothing. But I can't say the same about my mind which was busy acquiring things all day long for the first few days.

Looking back, I laugh at these early thoughts, for as the journey unfolded I forgot all about shopping, I had no time for it. In the end I managed only one hour of shopping -- on my last day. In addition to a few gifts for those at home, for myself I bought only a mini-prayer wheel for my altar, some prayer flags, and incense. No Buddhas, no kiras.

Don't worry dear reader, I'm not going to take you on a moment-by-moment description of my trip through Bhutan, we are just hitting a few high and low points here, but I do want to tell you about the next place we went.

Five years ago my dearest friend Katie cut out an article from the Sunday New York Times for me. It was about a valley in Bhutan where a large bird came to winter, the Black-necked crane. The people in the village consider them very auspicious birds. They call them messengers to heaven. Prayers are said to ensure the birds' safe return every year.

The article went on to describe tall poles with white cloth banners (flags for the dead) that are erected by a family in the year after someone dies. The wind catches the cloth, and each time it ripples another bad deed/sin from the deceased one's life is washed away. The resulting purity secures a good after-life for the family member. The article was very interesting, but the place was a remote valley, in a remote nation, and to see the birds one must be there at a specific time of year. When I read the article I never dreamed I would go there, but as we arrived at our hotel and I listened to my guide's description, I realized I was in that very place! Phobjikha! Elevation ten thousand feet, and at the perfect time of year to see the cranes!

My back was better so we took our first warm-up hike. It blew my mind, really, I could feel myself reading that newspaper clipping and thinking, *cool*. And now here I was, in the very scene I imagined, complete with flags and birds. *It felt like a miracle.* And in that very moment a red dragonfly landed on the ground in front of me – the first one I had seen since stepping out of the airport. Why hello, messenger from heaven.

After a few days of touring and altitude adjustment at last it was time to start the trek. We drove to the pass where I put on my pack and adjusted my hiking poles. Then I took a step, and then another. After all of my vacillation and worries and prayers I was doing it! I was on the TBT, and it felt good.

At first the trail took us through fields with long-haired, long-horned yaks. Some of them looked at us suspiciously. My guide said that sometimes they can get 'testy,' so I prepared to use my hiking poles to fend them off if necessary. Other than some very minor yak trepidation the trail was a dream: hills on both sides, rock-lined streams running though small valleys, large birds like hawks and griffins flying overhead, sunny blue skies and cool mountain-air breezes.

What had I been worried about? My body, praise her, was doing just fine. In the afternoon the trail took us between two closely spaced houses and then we entered the central square of a small village, Rukubji. Even here, in this place off the road and the usual routes, many houses had carved, complex, woodwork painted with intricate decorations. Small boys were playing a sort of hide and seek. Paths threaded through the village, between the houses. We came out at the village temple where we met our young driver. He spun all the prayer wheels at the temple, and I did too, right behind him. I wondered what he was praying for. For me, with each spin, it was *gratitude, gratitude, gratitude.*

We walked down a small hillside to get to the home where we would stay for the night. From the temple to the home the hillside was covered in ancient oak trees. This was an old-growth grove that was fiercely protected by the village. Even to break a twig off one of these trees meant you had to make amends to the whole village.

Our hosts had a porch where I removed my boots while admiring the ancient trees. Inside the house was a very plain central room with a primitive woodstove and a huge pot of boiling water on top of it. In the corner near the stove was an old woman smiling and spinning a hand-held prayer wheel. I was invited to sit next to her and given a mug of sweet tea.

There were a few other rooms with doors off this central room, one was the room where I would be sleeping – very basic with two single thin foam mattresses on the floor and a pile of blankets nearby. The house had electricity, but in my room there was just a single bare lightbulb overhead. No clock, no table lamp, no rugs, just a bare wood floor. The primitive kitchen had concrete floors and counters. The bathroom was also very rustic, and the concrete floor was wet. (I was coming to realize that in Bhutan one takes their shoes off outside, but then there is another pair of rubber sandals that you slip on before you enter the bathroom because there is no such thing as a shower stall or a tub in the bathroom – water goes everywhere.) To get water from the sink you had to turn the knob underneath – the faucet didn't work. There was a western style toilet, but I never figured out how to flush it.

But in this basic dwelling there was another door, and this door led to the temple room. The difference between the temple room and the other rooms was mind-blowing. In the temple room the colorful walls were hand painted with intricate patterns that seemed so perfect I thought it was wallpaper. When I asked, I was told no, it was all hand painted. On one side of the room was a huge altar, perhaps ten feet long by five feet wide. It was completely carved and painted and had many alcoves that contained statues of buddhas and deities, and handwritten scriptures. There was a butter lamp burning on the altar, and incense burning on a pan beneath it. There was a row of small metal bowls filled with water. The altar also held a mysterious metal pitcher with peacock feathers sticking out the top. It was hard to take it all in, but the overall impression was how much space, work, and wealth, went into creating this room -- and the other rooms, not so much.

There was a gathering that evening for one of the local schoolteachers who was going away. Ten people sat on the floor around the wood burning stove in one room, and ten people sat around an electric heater in the other room. There was tea, crunchy corn, barley snacks, and rice wine being offered around. My guide told me that the big topic of conversation was the TIGERS: an older pair, and two two-year old cubs, who were roaming near the village and occasionally killing a cow. One of the cubs had been collared, and there were camera traps installed along the roads. A few people had seen the tigers along the very route that we were planning to hike the next morning. My guide and I thought, *how cool*! But the partiers thought we were crazy and shouldn't go. "It's risky, madam, it's risky," said one of the male teachers dressed in his formal *gho* (plaid robe). For a minute I thought he was offering me whiskey, and I shook my head no.

After a lively conversation, that I couldn't understand a word of, it was decided that we would hike on that section of the trail the next day, but we would bring a local guide with us.

When the main meal was ready, dish after dish came out: red rice, spiced chicken, noodles with mushrooms, green beans, steamed pumpkin, chicken soup, broccoli and cauliflower, red chilis with cheese and garlic, lentil stew, and few more dishes I can't remember. What I do remember is that when all these dishes were set on the floor by the woodstove I was called to serve myself first, as the guest of honor. Everyone was sitting in the big circle, and I had to step to the center, crouch down, accept the offered plate and start spooning things onto it. There was background murmur and discussion (in a foreign language). Every eye was on me and I'm sure I made many mistakes both social and culinary. *She spooned the lentils onto her rice instead of into a bowl! There is no way she will be able to eat those chilis!* And on and on. I was fine with being a source of entertainment, it was my meager contribution.

The next day we made it through the valley of the tigers. We could hear them and see their tracks, but I was rushed through by the local guide, and I never got a glimpse of them.

At the end of that day's hike we arrived at another small village. Just imagine, you look up in those foreign windows, with faces peering out and gray smoke curling from the stovepipe, and you wonder what life is like in there; and then you are escorted inside. This is where you will be staying!

Every day we hiked, and every day, at the end of the day, I was feeling fine. One particular day we were doing a fairly short hike starting at a very high altitude, but the next day, I was warned, would be a very long, very rocky, hike.

The top of the pass was over twelve thousand feet in altitude, and the stream next to the trail was frozen. As I hefted my pack and adjusted my trekking poles my breath came out in steamy white clouds. The tops of the tall conifers were coated with ice crystals, as were the lichens draping down from the trees. When the lichen crystals caught the sunlight they reflected every color, like a prism. It was breathtaking – two days before Christmas and I had nature's tinsel and lights. Even the small plants near the ground were coated with crystalline ice. It was invigorating starting out in this chilly, high altitude, scene. Step by slippery step up we went, over the top of the pass, and then gradually heading down. After almost an hour of hiking down a trail lined with sparkling conifers my guide invited me to walk on by myself while he 'used the bush' and had a cigarette.

I enjoyed this alone time on the trail every day, either walking slowly in the silence, or sitting and meditating. On this day I walked for a while, but then decided to sit and rest on a low mossy log. The moment I sat I realized my mistake – the moss was saturated with water-- and I sprang up. But my left knee was not ready for that extreme, sudden, movement, and I had that sharp, sudden, sickening, pain of a 'tweaked' knee.

Shit, shit, shit. Of all times and all places for my sturdy knee to decide it had had enough. What now? Standing and leaning on my poles I shook my leg trying to put things back in alignment. I wasn't sure what would happen when I tried to put weight on it, but when I did I was pleasantly surprised that it didn't bother me much. I was able to continue the hike.

That night in bed the knee started hurting, and we were supposed to start our longest, most difficult hike yet, at six-thirty in the morning! Would I be doing permanent harm to myself? The type that would require surgery? If I went would I forever regret it? If I didn't go would I regret that? After all, I wanted to be challenged, that's why I was here, and the trail was supposed to be really cool and lined with old-growth forest. *What to do? What to do?* There was only one thing to do – pray. I wasn't praying for a decision so much as I was praying that whatever decision I made was the *right* decision. But first I needed to pray for healing.

If you believe in healing energies, as I do, well why not use them on yourself? Why does it feel almost embarrassing to attempt self-healing in this way? Even now, I must make a conscious intention and overcome psychological barriers to laying on hands for healing. But I wanted to be healed, and I had nothing to lose, so I did. I moved my hands to my knee and I silently sent energy there and prayed for healing. That's all I could do. My decision would come in the morning. And I drifted into a deep sleep.

In the morning my knee felt fine! No pain at all...but I knew that a rocky ten-mile downhill hike might change all that. We got a ride to the summit, put on packs, grabbed the poles, and I headed out, for better or for worse. There would be no possibility of a ride from now until we got to the bottom. I was told the hike would take about six hours.

At first I cautiously picked smaller steps and turned to favor my strong side when there was a long drop. I was aware that at any moment things could go really wrong, then I'd just have to depend on the Advil I was carrying to cure the pain and forget about any more trekking on this trip. But step after step- mile after rocky, slippery, mile my knee held out, and there was no pain whatsoever. My mantra became *gratitude, gratitude, gratitude* down the long decline while I enjoyed looking up at the huge old trees covered with mosses and lichens and orchids.

Lest you think this is some sort of just-so story, not only did the knee start hurting again, but my other knee began hurting too, then my hips, then the tips of my toes. After many, many, miles everything hurt. Now I would need not just ACL surgery but a hip replacement too. So many things hurt it wasn't worth trying to fix any of them. And there I was, sliding down the slopes as if I were skiing. Blisters were forming on the inner margins of my thumbs from the trekking poles, but I could not do without them, they were saving my life every dozen steps or so. And on and on and ON it went.

The sun set behind the mountain. It was ridiculous to think anyone could hike this far in one day, we had been at this for more than eight hours now. It was Christmas Eve. My guide's friends were expecting us for dinner, but we were still on the trail. I didn't complain, what good would that do? We just had to get down. That's all there was to it, so on we went. Then at last, at last, there was the bridge across the river! A long foot-only suspension bridge. The guide gave me a high five and said "we're almost there now." A short path to the road, I thought, and our car would be waiting. Relief at last!

At the end of the suspension bridge there was a T where one could turn right or left. I jokingly said "eeenie, meenie, minie, moe," pointing left, right, left, right. And my guide said, "this way," so we made the left. And we walked, and walked, and walked.

Soon it was clear that we were not almost there. One indicator was that my guide was walking faster and faster. I'm convinced this is a Y-chromosome linked trait: when men are lost and unsure, instead of stopping or slowing down or gathering data, they push on even more quickly. My guide who had been so kind and protective all along had now left me behind in the dark on a narrow jungle trail along a river.

I stayed calm, even though I knew we were lost and he wouldn't admit it. I had a headlamp in my pack and I stopped to get it out. Soon the trail dissolved into mud, and next thing I knew I was climbing a waterfall in the dark, hand over hand.

"Maybe we should check the map," I suggested.

"No, this trail has got to lead somewhere," the guide replied. "See that water pipe? It is leading to a house."

He had nothing to say a mile later when we saw the pipe emptying into the river. I didn't say anything either. Farther and farther and farther we went, and darker and darker it got. I was tired, sore, and hungry, but I silently vowed not to question or complain. I'm sure he was feeling rough too, and with an added layer of embarrassment. In fact, the few times he stopped to wait for me, I greeted him with a calm expression and a smile. In the tiny booklet that advises the Bhutanese people how to behave one is instructed to walk "calmly and quietly like a cat."[21] And that was my practice.

When it was very late, and fully dark, the trail turned into a machete swath for one, made a few weeks ago by the looks of it.

The guide's mobile phone kept ringing and ringing. "*Kozombola* (Hello)," he answered. The calls alternated between our dinner hosts and our driver. I wondered what he was telling them. Was he telling them we were lost? (We most definitely were.) Or was he telling them that the American lady was going very slowly and that's why we were late? (Not true). I doubted he was telling them we were lost. He asked the driver to turn on his lights so we could see the car. Then he showed me a light far up ahead on what I assumed was a road along the riverbank. "See, that's the car." But as we hiked closer and neared the light I could see it was coming from a small farmhouse and not from a car at all. As we approached the farmhouse he took down the round wooden timbers that were creating a rudimentary gate, and he motioned me through. I wasn't afraid of the tigers, but at this point I became frightened. What was in that pasture? A bull? A protective dog? We were definitely invading this space, unexpected, unprotected, and in darkness. Fortunately, nothing gored me or bit me, and the kind farmer who lived there pulled on his boots, grabbed a flashlight, and hiked with us out to a small dirt road where our driver could find us.

I never complained, but I did inform my guide that the next day, Christmas Day, I would not be hiking *anywhere.* As it turned out, it was the only day I took 'off,' and by December 26th I was completely recovered and ready to go.

Part of my mission on this trek was to see if there were any old-growth forests in Bhutan. My research informed me that about seventy percent of the nation is covered with forest, and they have made a commitment to never let that figure drop beneath sixty-two percent, but I could find no information anywhere on the amount of original forest left, if any. This should not be too surprising because even in the United States we still have no national database on the locations of our old-growth forests.[22] Day after day, wherever we went, trekking, or by car, we were on the search for ancient forests. And we found so many of them! Mostly on ridgetops, but also along the trails. Impressively large trees that I had to stop and admire and photograph. It was so healing to spend most of my days outdoors hiking through forests!

Because the country had reopened after Covid just ten weeks ago, because this was a new trail, and because it was the off season (most people visit in spring or fall), we didn't see a single other visitor on the trail. Most days we saw no one on the trail at all, even locals. One day we did see a young woman carrying vegetables to feed to her cows. She was wearing rubber boots and her hand-woven basket, filled with broccoli, cauliflower, and cabbage, was supported by a strap across her forehead. It looked like a scene from an old wood-cut, until she got closer and we could see her white Apple earbuds. Hey, Bhutanese farm girls like music and podcasts too!

The weather was ideal for hiking at this altitude: not too cold, not too hot, and it never rained. Some days the trail was easy enough that I didn't have to watch every step and I could relax and look around.

One day it felt like I was walking in place and the whole planet and all her mountains were moving past me. It brought to mind the Mountains and Water Sutra, written by a Japanese Buddhist monk in 1240 which says (in part), "the green mountains are always walking do not doubt mountains' walking even though it does not look the same as human walking. If you doubt mountains' walking, you do not know your own walking; when your understanding is shallow you doubt the phrase 'green mountains are walking,' because you are just looking through a bamboo tube at a corner of the sky." [23] For one magical moment I thought I truly knew what that sutra meant.

As Phil Cousineau writes in *The Art of Pilgrimage,* "We learn by going where we have to go; we arrive when we find ourselves on the road walking toward us." I had arrived.

At last I reached the end of my trek on the Trans Bhutan Trail. My knees held up, as did my hips. My back is without pain. In all those dozens and dozens of miles walking over rocky terrain I had not fallen. I consider this all a miracle. Can it be attributed to my turning of the prayer wheels? Or to my powerful guardian angel(s)? Or, more mundanely, to my level of fitness? I don't have the answer to these questions, but I am grateful, nonetheless.

Although I was finished with the trail I still had two stops left – a stay at a monastery and the Tiger's Nest.

Fortunately, the woman who arranged my tour had both a brother and a brother-in-law who were reincarnates (Tulkus). They invited me to stay at their monastery. An important festival was scheduled to take place when I was there.

The day we arrived, vendors were set up outside the temple selling incense, coats, shoes, and plastic toys, among other things. The sounds of cymbals and horns and chanting was coming from the temple. We stopped to purchase an inexpensive silk scarf (chadra). At the appropriate moment, my guide and I were directed down the center aisle of the impressive temple, and straight to the Master of Ceremonies—the Rinpoche. As is typical, he took the scarf from my two hands into his, and then he handed it back, blessed. With the formalities over, I was led to a spot where I could sit and stare.

Up front the grand temple had a massive golden statue of the Buddha. The center of the space was filled with a hundred monks in maroon robes reciting teachings in unison and playing horns, cymbals, and drums. The monks were sitting on cushions, because they would be reciting for days straight with only short breaks, but everyone else, including me, was sitting cross-legged on the floor. Hundreds of local people – many spinning small prayer wheels or fingering mala beads -- were sitting all around me. I was the only 'westerner' in the place, and maybe because of that, in addition to my age, I was considered the guest of honor. When we broke for meals, everyone insisted that I go first. When tea was served during the ceremonies it came to me first. I cooperated, but deep down I knew that I was likely the least important of all the people there.

As I drifted to sleep in my room that night I could hear the monks still playing their instruments and chanting.

The morning of New Year's Eve we made our way from the eastern part of the country back to the west. A single short flight retraced all the miles I had just walked. Young people were partying in the town of Paro, where the airport is located, but I went to bed early to be ready for the Tiger's Nest the next morning.

Although we got an early start, some were even earlier and they were already on their way down as we were headed up. As we passed them we would greet each with a nod and a smile, 'Happy New Year.' 'Happy New Year.' 'Happy New Year.' At least a hundred times, Like so many chirping birds.

Our destination was that many-storied place built high up on impossible cliffs. Its exact story is complex and ever changing, but my version is that one of the earliest Buddhists from India, Guru Rinpoche (also known as the lotus-born Padmasambhava), traveled through the Himalayan mountains to teach tantric Buddhism. He carried daggers and subdued the local spirits, he was fearful looking. He had two female consorts with him as he traveled in Bhutan, and it is said that he transformed one of them into a flying tigress to deliver him to the cave in the mountainside that is now the heart of the Tiger's Nest. Some Bhutanese take this literally, but I think of it as a metaphor illustrating that he had so much personal power he could cut through any obstacles, be they wild animals, harsh weather, combative landlords, or steep mountainsides. Or maybe it was his consort, the tigress herself, who had this power to clear obstacles.

Guru Rinpoche spent three months meditating in the small cave in the cliff face (others claim it was three years, three months, three weeks, three days, and three hours), and when he emerged he was a peaceful Buddha. Today Guru Rinpoche is the primary spiritual figure worshipped in Bhutan. Three more temples were added on to the original one surrounding the cave; the first was built in 1508, the second was built in 1694, and additions were made in 1865 and 1983. It's difficult to imagine how they hauled all those building materials up the side of those steep rock cliffs.

After a few hours of hiking straight up we reached the Tiger's Nest. We left our shoes and packs outside, including our cell phones – no photographs allowed. Another guide, a friend of my guide, offered to watch them.

I entered the first temple, a tiny place, and on my left, I could see the small, closed door that led to the cave where Rinpoche had meditated. The door is only opened once a year. The temple altar had the usual bowls of water and the golden statues. My guide and I prostrated before the altar. When I rose, a monk blessed me with water from one of the peacock feather vessels – said to remove all negative energies.

Then we walked the steps and pathways to the second temple. My guide knew I liked to meditate in such spaces, and he pointed out a corner where I could sit. Very quickly I 'dropped in' to that other state of consciousness where day-to-day thoughts have no power. I often sense it as emptiness. But what is the emptiness like here? Is it different? It seems to be different, but I can't call up any shapes or colors or words to say how it is different. I seemed to be experiencing not just 'empty-mind' but 'God-mind.' After about half an hour I got up in a daze, put a monetary offering on the altar, and turned for a blessing from the abbot of the temple. He touched my head with a gold object. On the way out I lit butter candles for Rick, for our daughter, for my grandson, and for myself.

I had had a dream, and at times I wondered if I had reached too far for that brass ring, but here I was, in this moment, stepping out of the mystical Tiger's Nest. The golden doorways were behind me and in front of me was a majestic valley. I had done it. I had trekked for over a hundred miles across Bhutan, I had stayed healthy, I had made it to the Tiger's Nest. The wave of emotion was overwhelming -- it was mostly gratitude mixed with relief, but there were other feelings too. Rick was there in some way. And then the tears came. A dream. A dream come true. My life. This moment. Tears of sorrow and tears of joy. I was here, here, here. On all planes of existence. I rested my head on the stone wall and let the tears flow and my shoulders shake.

"Madam," said the kind guide who was watching over my pack, "it will be alright."

Yes, yes it will be.

It will be more than alright. I am a wild woman, full of adventure, and deeply connected to both the earth and the divine. I had seen langurs, both gray and golden, I had seen red pandas, I had seen black-necked cranes – the bird of heaven – I had heard tigers, I had walked through pastures of yaks and Mongolian horses, I had seen many huge trees, I had seen hundreds of thousands of prayer flags flapping in the wind, I had walked day after day. I had stayed in a monastery and had a private audience with the Rinpoche. I had meditated in temples and altar rooms of every kind (but my favorite space was still in my bed – wherever that happened to be). I had slept in the Royal Suite – literally outfitted for a Queen; but I had also slept in my hiking clothes on the floor of an empty truck-stop. I had drunk cup after cup of the sweet tea that was offered to me everywhere I went. I was lost in the forest – in the dark – with a smile on my face. I had prayed, and my prayers had been answered.

When my daughter and nine-year-old grandson came to pick me up from the airport he asked me, "Gigi, why did you go, anyway?" He had missed me. The answer that came to me was simply: "adventure." I had taken a chance, I had been somewhere very, very, new. I had stretched myself; I had been on a wild adventure and I lived to tell about it. When I woke up in my own blessed bed I thought of the Shackleton explorers. Eventually they, too, woke up in their beds at home. What that must have felt like! *Home, sweet, home.* I returned stronger than when I left.

It had been a wavespell of years since Rick passed over. What a wave.

Meta One

Starting a book takes a certain kind of confidence: first you must imagine you have the perseverance to finish it; then you must believe that it will be useful or good enough to find a publisher or an audience, or hopefully, both.

Writing the memoir felt good, it felt risky, but I was ready to face to world with my wild truth. The only trouble was that when I stepped back to read it with fresh eyes it didn't seem to be the fabulous book I imagined it would be. I had imagined a memoir that would bring in some income for this last part of my life, but ... it wasn't that book.

Although it *was* a book, and didn't it deserve 'life' too? (Life in this case meaning publication.)

Writing the memoir made me want to write more, but first I had to do what I could with this manuscript. I started removing things, beginning with all the 'advice' for those entering older age: advice about downsizing, advice about travel, advice about money, advice about health, advice about retirement. This is not a self-help book, I told myself. I kept going and removed a section on meditation. The relationship advice was the last to go, but finally I removed most of that too. The memoir was more concise now, but in my eyes it still wasn't great. Should I just abandon the whole thing? I was stuck. What could move me forward?

When I wrote my previous five books, I never shared anything before publication, except with my publisher and the readers they sent it to. I was so afraid criticism would burst my bubble of creativity that I held my words close. But this time things were different. I needed help and I knew it. Sure, I could have dished out thousands of dollars for online workshops or writers' conferences. But who knew if the teachers were any good or if they'd just dull the blades of my words?

I needed to get unstuck, so I worked up the courage to share with others, both writers and readers.

My first big share was with my friend, a poet and creative writing teacher. She had just broken a bone and was not able to move around much. I figured it was a good time for her to read. She was the perfect choice, so serious, yet so gentle at the same time. She thought it had great potential, but it needed an additional voice that addressed the readers directly. Her observation rang true to me and in response I added italicized interjections. That helped, as the interjections allowed me a dreamy voice, somewhat removed from the story I was telling. It strengthened the book, but it also confused some of the readers who came next. I ended up changing the font to regular but leaving the words in.

Then I asked the question I would eventually ask others: "What's the weakest part?" I recognize that it is a difficult question to answer, for a friend anyway. She suggested that I leave out the story of Peter Pan arriving in Mexico for the first time and trying to buy twenty dollars' worth of pot which, as it turned out, was a whole bale. It was a good story, very funny, and I thought I told it well, but it was not useful to the memoir. When she pointed that out I could see it right away. Only the story. Stick to the story. As writer Natalie Goldberg says, "Don't be distracted by the flies."[24]

My friend also suggested that I might want to explain myself to my current fans – the 'tree people' who had read my first books. She was honestly, yet respectfully, addressing the elephant…. I had made myself very vulnerable by writing about drugs, sex, jealousy, and hate. Some who had respect for me before may have it no longer.

I tried to appease my fellow tree people, but it never felt very smooth – apparently, I had one voice when I talked about trees and another voice when I talked about more personal things. I cut many of the forest sections too. Besides, I was still working to save trees and forests every day, it was my life. I was flying on planes to stand on stages. I was attending local planning board meetings. I was spending time with the trees themselves. I didn't feel like I had to apologize or explain to anyone.

Another writer friend with deep credentials – someone who gets paid for coaching writers, but who kindly read my manuscript for free – also wanted more *botany* in the book. I looked around for where that might go, but, again, the trees are a different story. Overall, however, she loved the book, and felt my voice was "warm and true," (how kind!) and that it had commercial potential.

Next, I shared it with a reader friend, this woman has great taste in books, and she has read more than anyone I know. She thought it had commercial potential, but she didn't like the parts where I sounded mean and vindictive when I talked about Jerry, or the chapter where I described the evil county executive who had ordered the teardown of my farmhouse. Honestly, I was a little uncomfortable with those parts too. Was I using the book as an opportunity to get even, although it distracted from the story? Maybe.

I removed the story of my house being torn down in retaliation for my forest-saving work (perhaps an essay someday?), and then slowly, somewhat reluctantly, removed some of the harsher phrases from the Jerry chapter.

I was further encouraged to soften that chapter in response to comments from my daughter. She simply said, "Are you sure you want people to know all that?"

I was on a roll with the sharing now. I wasn't embarrassed anymore. In addition to the suggestions I had asked for, most readers really liked it overall and thought it would be popular.

Next, I bravely sent it to a publisher who had recently interviewed me on a podcast. They were interested, but then I hesitated. Was I selling it short? With that publishing house it wouldn't get much distribution. And then there was the bigger hesitation...what would friends, family, and former lovers feel about their appearance in the book (even if somewhat disguised?).

Jerry wasn't interested in reading the manuscript, and he had already told me he didn't care what I had to say. "I can handle it," he said.
Next, I reached out to the redwood-wedding couple and the Mexican wizard. Both were fine with what I had shared.
Then for a week the manuscript sat idle.
I believe things have a certain energy, even nonliving things like ideas and essays, and when the energy leaves them they may 'die.' In other words, they become irrelevant, they become like my friend's novels sitting under his bed, never to be shared. I wanted to keep the memoir alive. Keep the energy moving.
On a whim I signed up for a free online webinar on self-publishing. The presenter stressed that if you give your book to a publisher you might earn a

dollar a copy in royalties, but if you publish yourself you could earn fifteen dollars a copy. He said it's easy these days with print-on demand. The next day I dove into the Amazon self-publishing process and by the end of the day I had ordered author's copies. A few days later they came in the mail. There was my paperback book in hand!

I still considered it a draft, and it wasn't available to the public, but going from a computer file to a physical object was a big step.

To publish or not? To self-publish or not? These questions were on my mind daily. I reminded myself that I had barely reached out to a few agents and presses. Was I selling myself short by going directly through Amazon?

It was the perfect time to read Rick Rubin's book about creativity.[25] He says to let the project go so the next thing can come on. I interpreted letting it go as releasing it into the world. It echoed what I told my daughter in response to her question. I told her that writing was my creative work, and to keep the creativity flowing I had to honor and move along each thing before I could get to the next.

Should a painter complete a painting, even if they realize halfway through that it will not be their favorite? Only the artist can answer that. Not every painting can be a masterpiece, but perhaps it is meant to be someone else's favorite – even if not the artist's.

But then I hit another slow spot. What was going on now? What was my current obstacle? It was my mother. She was ninety-two, and in an independent living apartment. I didn't want anything I wrote to hurt her – from sex to drugs to my childhood wounding. It was not worth it to me to disrupt our relationship for the sake of the book.

I gathered my courage and put a copy of the book in the mail to her with a note explaining that it was just a draft, and anything could still be changed if it bothered her. Two days later one of my sisters contacted me and said that mom had fallen and was in the hospital. For a moment I worried that she had already gotten the book, was disturbed by what she read, and that had somehow precipitated a fall. But, no, the book was still in her mailbox, unopened.

My sister brought mom's mail to the hospital the day I arrived to sit bedside. In between her hallucinations brought on by pain medication – including an Elvis sighting – the hours went by slowly. There sat the book, so I offered to read it to her.

She loved the reading! And it was a huge gift for me too, to be able to see her reaction to each word. Together we cried about Rick – she loved him too – and together we laughed about the guy trying to squeeze his junk through a plastic ring. She understood my relationship with Jerry for the first time, and worried about my upcoming trip to Colorado. Overall, it was bonding, not divisive, and she thanked me again and again for reading it to her. It was one small ray of light during those pain-filled, fluorescent-tinged days. She couldn't stop talking about "Joan's new book."

Meta Two

The book was getting better and better, and my worries about it were fewer.

I had a conservation meeting coming up in Colorado. Jerry would be there, and I was looking forward to seeing him. I had changed the chapter title from Love/Hate to Love/Hate/Love and handed him the draft with a little speech about how much I appreciated the freedom to be able to write what I needed to write with no push-back. And it's true, this book would have never existed without that. If he had been some ego-driven high-powered lawyer I wouldn't have risked the possible lawsuits from publication.

For our first three days in Colorado we sat together in meditation every morning at 6:30 and just generally hung out with each other for the rest of the day. We were getting along great, so great, in fact, that someone from the group – who knew all about our history – got up the courage at the final dinner to ask why we weren't still together. And yes, there was alcohol involved.

We could have laughed off the question, or deflected it, but instead Jerry started telling a story about our very early days when I went off and had a 'date' with another man.

Why was he telling that story? Why was he purposely trying to make me look bad? I jumped in to defend myself by telling the full story – that I had been perfectly honest with him about what my plans were, and I had told him I wouldn't go if he didn't want me too. (In essence I was asking *how much do you like me? How serious are we? Because if you are serious about us, I won't go.*) I wanted him to ask me to stay, but he acted like he didn't care if I stayed or left. So, I left. And now, ten years later, he was doing this?

Then other stories came tumbling out. It didn't take long for the others at the table to see perfectly well why we weren't still a couple -- especially after he called me *a selfish bitch*. Whew. I had to get up from the table at that one. I had never been called anything like that in my life.

I flew home the next day never wanting to see him again. When would I ever learn? Well maybe this was the time I could finally end it. He had been so unkind, and I had witnesses.

My spiritual books tell me I should remain detached from the anger that arises from such pokes. I should merely witness his behavior and my reaction to it, then let it go. But I wasn't there yet.

He claimed later that he was describing how he felt ten years ago. But even so, if he felt that way then he was hiding it. Like he hid so many things.

There are other names I would claim, and I'm sure at times I could be a bitch, but selfish I am not. Likely the day he had that thought I was cooking meals for him, dressing for him, and sharing a bed with him.

Then he changed his story again. "It was a sarcastic joke," he said. "I was trying to lighten the mood."

"No one was laughing," I replied.

Then his story changed again. "I was trying to get even for the mean things you said in the book."

"How did you know what was in the book? I had just given it to you, and you hadn't read it yet."

"Yea, well, I could imagine."

How can I believe anything he says? How can I get to the root of the problem if it keeps moving? I still don't know the truth behind his mean comment. The only thing that makes sense to me is that he is continually looking for my soft spots and poking them. I don't know what motivates him to do that, but I have given up trying to figure it out.

After I headed home from Colorado the days of silence prompted him to pick up the book and read it. He emailed that he wanted to discuss it with me. This was an invitation I couldn't resist. Maybe it would be our last conversation, but I needed to hear what he had to say. Besides, I was softening a little at that point. I had given deep thought to what life without him would be like, and I realized that although I had many, many acquaintances and casual friends, I only had one other close friend, Kali. But she was a recluse, and rarely ever wanted to visit.

I was aging and, according to the sociological studies, I needed friends –Maybe I couldn't afford to shut Jerry out, even if he was cruel.

I reached out to Kali and told her I needed to see her right away – something I never did. Bless her, she came, listened to my story about my latest trip to Colorado (she had heard many Jerry stories by now), and she thought that I should forgive him. Her husband was also prone to cruel remarks on occasion, she said, and men are just stupid and will eventually regret their stupid remarks.

When it was time for her to go home, I sent her with a paperback copy of the draft to read. I decided to answer the phone the next time Jerry called.

He started the call by being humble and apologizing about his remark at the meeting. I was listening closely; it was the first time I ever heard him genuinely apologize for a specific action.

I was holding my breath when it came time to discuss the book, but he only objected to one word. That particular word bothered him so much that I easily agreed to take it out. I no longer had to worry about the rest. Phew. Another Jerry roller-coaster come and gone leaving us wondering what had just happened and what would happen next.

I made the small change he asked for, and a few others, but maybe this book was not finished after all? I had finished five books previously, and I knew exactly when they were done, but this one was my life, and it was still ongoing. The lessons were still coming. And I kept writing.

It started to feel like <u>this</u> was the real story -- the book after the book. The sharing, the reactions, the hesitations, this was the meta layer no one talks about. It began to feel like jazz. Slow jazz. Small changes and additions would bubble up at random times during the day. This was nothing like sitting down to a thousand words. It was more like waiting for a magma bubble to rise from the center of the earth, or a piece of stardust to fall from the sky.

And it was not over with Jerry. His objections were building. He gave up trying to convince me I was ruining my reputation, and then he switched to worrying about his own.

I thought of what writer Anne Lamott said: *You own everything that happened to you. Tell your stories. If people wanted you to write warmly about them, they should have behaved better.* [26]

I didn't share that passage with Jerry; he was having a hard enough time as it was. Someone who wanted to think he didn't have much of an ego was now defending it mightily. I reassured him that I was still working on the manuscript, and I did plan to soften it somewhat.

Then I got an email from him asking me to burn the chapter. I agreed to a call, which I was not looking forward to, since I knew it would be more diatribe than dialogue. I started by telling him (nicely) that I was not going to burn the chapter. I agreed that it was not pleasant, but the pain from that time was a significant part of my life's story.

His next ask, if I wasn't going to burn the chapter, was that I should end it on a positive note. I should tell the readers that this is how I felt in the past, but now we have survived that and come around to love each other. He wanted a nice bow on it.

But, I reminded him, that's not how it is, to tell the story otherwise would be dishonest. *We are still on a see-saw* I started to say, but the words *teeter-totter* came out instead.

Strangely, I ran into those same words three more times in as many days. The universe gives us signs.

Jerry wasn't the only one who wanted me to burn that chapter. My friend Kali had finished the draft I had given her, and she thought I should leave the chapter out too. She said it turned the book from a nice pleasant read into something she didn't enjoy for a number of pages.

She also reconsidered her advice about forgiving him.

It *was* a difficult chapter. I couldn't remove it, but I took out more of the darker things he did, and the darker thoughts I had.

I asked Jerry if he wanted to help rewrite the section on his childhood trauma, as he had offered. It was a small kindness, and I thought it might help, but what I got back was a spitting-mad, pain-filled reply. I could understand that, but in stepping back I noted that his objections were all about other's possible judgements of us, and he never addressed his damaging actions or my heartbreak. He still had excuses for everything he did.

It was his inconsistency, really, that was getting to me, one minute sweet and the next bitter. And now even the small things I would have overlooked in another relationship – the minor fib, the teasing put-down -- felt like the perpetual rescraping of a wound that had not quite healed. I reacted sharply to the pain it brought on. We were both hurting, and hurting each other.

As I tried to sleep that night the words *trauma bond* jumped into my mind. I wasn't even sure what they meant. Is that what was happening? Two people trapped in a spiral? At 2am I was scrolling through podcasts on trauma bonding.

I learned that it was more of a one-way thing. Someone alternates being kind with being unkind, and the bondee sticks around waiting for the next dose of kindness-sugar. It's the intermittent reward that creates the bond, and according to the best podcast I found, the only way out was *no contact.*

To help with taking the big step of no contact, they say you cannot view your partner as complex, as someone to feel sorry for because of their wounding, you must only look at their hurtful side. And while that may be good advice for a woman in her twenties, I wondered if it was the best advice for a woman of my age.

Was it cowardly to want out of this pain cycle? To just want peace. Maybe there was more growth in going *through* it? Holy teacher Nisargadatta said, "the mind creates the abyss, the heart crosses it." Was I able to get across the abyss?

I was closer to Jerry than to almost anyone else. We knew each other so well. And beyond what we knew of each other on this physical plane – as messy as that was – there is another realm, a spirit realm, where we are connected, where we are each learning life lessons important to this incarnation. Our dance of love and hate is just a small part of who we really are. We are all seeking divine connection. But that's the aerial view.

We are also beings in soft human bodies. Our chests rise and fall with each breath. My body felt ready for kindness, for peace.

I could see how the book was a catalyst. It's likely I wrote that chapter because I wanted him to deeply understand what he did and what I felt. Difficult 'conversations' are easier for me in writing. In black and white I can take my time and make my case. It's the same way I felt when I was writing *Teaching the Trees* -- it was a story I was telling the foresters. That's what writers do, I suppose.

Although many wanted this difficult-relationship stuff out of the book, the drama kept going on, and instead of removing, I kept adding more in.

I recognized that it wasn't for the reader anymore, it was for me. I started to feel as if I were writing my way back to solid ground.

Post Script

Any story of our lives must have a frame around it. We cannot include every detail so we pick and choose. That is why some writers, such as Alice Walker, can write numerous memoirs – she uses different frames for different books. My frame here has been a certain time in my life, my most recent years, and I didn't realize until it was finished how much of the canvas would be taken up by men...specifically my body and its relation to men. My daughter and my girlfriends were a big part of my life during those years, but I hardly see them here at all.

Side by side with the stories of the men are the stories of my adventures in the world. I see these as woven from the same cloth for it takes bravery, doesn't it, to open oneself to the world of the *other*, whether man or woman or mountain. I see in these words that I am a brave one.

I never kept in touch with my Mexican guru Sri Juan. That was an era before we were connected by social media, and I don't even know his real name. Maybe my experience of feeling so beautiful on the beach that day was a reflection of his spirit more than anything. My short time with him remains a touchstone for me, and my spiritual heart has continued to open since that time.

I never saw the Wizard again either. I never really expected to. We follow each other's amazing life experiences through the medium of Facebook, and I believe both of us shelter a warm spot in our hearts for the other.

I am still in touch with Julian, we text weekly and have visited in person a few times, but between our distant homes and our personal responsibilities I don't see that relationship going deeper, as gorgeous and fun as that magical crystal man is.

I was invited back to Burning Man in 2023 but I chose not to go. I doubt if I'll ever go back, though I do continue to send a yearly donation to those who are building the Temple.

At the moment of this writing, Jerry and I continue to talk. I don't know where we will be by the time you are reading this. He still doesn't have a partner. Neither do I. There is no denying that we remain connected in some strange way. For better or for worse.

Five months after I returned from Bhutan I slipped on the bottom step in my house and badly sprained my ankle. After a trip to the emergency room, I had to wear a brace for weeks. The ankle was painful for months. I was so grateful it hadn't happened before or during my journey.

What have I learned? I can do hard things. Hard things are everywhere, but so is grace. I live alone. I love my life. It will be over before I know it. As I look overhead I see the trees that share this beautiful planet with me. As I look back I see the footsteps I will never walk again, as I look down I see only foam trails in the sea, as I look far out I see my grandson surfing the waves.

And now the book is finished.

Questions for Reading Groups

1. Do you think the author should break up with Jerry? If so, should she go 'no contact' or should she remain friends with him?
2. Should she have broken up with Jerry sooner?
3. Has the author negatively affected her reputation by publishing this book? If so, does that matter?
4. Have you ever experienced heartbreak? If so, do you think it is an experience everyone should have to become fully human?
5. Have you ever written, or considered writing, a memoir yourself?
6. Have you ever travelled to far-away places on your own, as this author has?

7. Are there places on earth where you feel strong energy from the land? If so, where?
8. Do you have experience using psychedelics or are you curious about trying them? Is this something you are comfortable discussing? Why or why not?
9. Have you ever been present at a human death? If so, was it was it horribly sad, somewhat magical, or something in between?
10. Have you ever had a dream-like premonition?
11. Do you have a "spirit animal" in your life (like the author's dragonfly)? If so, what is it?

Joan Maloof

is the author of five previous books: *Teaching the Trees; Among the Ancients, The Living Forest; Treepedia;* and *Nature's Temples.* She is the founder of the Old-Growth Forest Network, a national nonprofit organization. Maloof has a BS in Plant Science, and MS in Environmental Science, and a PhD in Ecology. She is a professor emerita at Salisbury University, and a fellow in the Explorer's Club.

If you enjoyed this book please write a short review on its Amazon page and recommend the book to others.

References

[1] Basho, Matsuo, *The Narrow Road to the Deep North* (Penguin, 2005)

[2] Freud, S, Dreaming and Telepathy, G. Devereux (ed.), *Psychoanalysis and the Occult* (New York: International Universities Press, 1953), p. 86.

[3] Ullman, Montague, Stanley Krippner and Alan Vaughn, *Dream Telepathy: Experiments in Nocturnal ESP, 2nd Ed.* (Jefferson, North Carolina: McFarland & company, Inc., 1989), p. 20.

[4] David W. Moore, "Three in Four Americans Believe in Paranormal: Little Change from Similar Results in 2001." (Gallup News Service, June 16, 2005). Retrieved on April 3, 2021. https://home.sandiego.edu/~baber/logic/gallup.ht ml

[5] "My Grandfather's Clock" is a song written in 1876 by Henry Clay Work

[6] Albert Einstein, letter of condolence to the family of Michele Besso (March 21, 1955), quoted in Tabatha Yeatts, Albert Einstein (New York: Sterling, 2007), p. 116

[7] Roper Center for Public Opinion Research, "Paradise Polled: Americans and the Afterlife." Retrieved on June 17, 2021. https://ropercenter.cornell.edu/paradise-polled-americans-and-afterlife.

[8] *The Essential Rumi*, ed. Coleman Barks. "Who says words with my mouth?" p. 2.

[9] Rilke, Rainer Maria, *Archaic Torso of Apollo*, 1875 –1926.

[10] Frankl, Vicktor E., *Man's Search for Meaning*, Beacon Press, reprinted 2014. Pg 35.

[11] "The King and the Handmaiden and the Doctor," in *The Essential Rumi*, translations by Coleman Barks. Castle Books, 1995, pg 229.

[12] Lewis, Thomas, Fari Amini and Richard Lannon. *A General Theory of Love*. Random House. 2000. pg 207.

[13] Hoagland, E. *The Courage of Turtles: Fifteen essays about compassion, pain, and love*. Lyons & Burford, 1993. pg 133.

[14] Pockett, Susan. *The Nature of Consciousness: A Hypothesis*. iUniverse, 2020.

[15] Dried resin from *Bursera* or *Protium copal* trees

[16] Zeppa, Jamie. *Beyond the Sky and the Earth : A Journey into Bhutan*. Riverhead Books, 1999. pg 174.

[17] Crossette, Barbara. *So Close to Heaven : The Vanishing Buddhist Kingdoms of the Himalayas*, Vintage Books, 1995. Pgs 10, 27, 47.

[18] Jalāl al-Dīn Rūmī . *The Essential Rumi*. Translated by Coleman Barks, Castle Books, 1997. pg 97.

[19] Potts, Mary Anne, "The 25 most adventurous women of the past 25 years", Wayback Machine, *Men's Journal*. October 1, 2020.

[20] Sarno, John E. *Healing Back Pain: The Mind Body Connection*. Grand Central Publishing, 2019.

[21] Yoezer, Jigme, *The Bhutanese Art of Harmonious Living*, 2020. pg 38.

[22] In April 2022, President Biden announced Executive Order 14072 calling for an inventory of old-growth and mature

forests on federal lands. This section was written in March 2023. Since that date various maps and databases have been developed.

[23] *From Moon in a Dewdrop: Writings of Zen Master Dogen* [Japan, 1200-1253], edited by Kazuaki Tanahashi.

[24] Goldberg, Natalie. *Writing Down the Bones*, Shambhala, 1986

[25] Rubin, Rick, *The Creative Act,* Penguin Press, 2023.

[26] Lamott, Anne. *Bird by Bird*, Pantheon Press, 1994.

Printed in Great Britain
by Amazon